Strengthening Design of Reinforced Concrete with FRP

Composite Materials: Analysis and Design

Series Editor

Ever J. Barbero

PUBLISHED

Strengthening Design of Reinforced Concrete with FRP, *Hayder A. Rasheed*

Smart Composites: Mechanics and Design, *Rani Elhajjar, Valeria La Saponara, and Anastasia Muliana*

Finite Element Analysis of Composite Materials Using ANSYS,® Second Edition, *Ever J. Barbero*

Finite Element Analysis of Composite Materials using Abaqus,™ *Ever J. Barbero*

FRP Deck and Steel Girder Bridge Systems: Analysis and Design, *Julio F. Davalos, An Chen, Bin Zou, and Pizhong Qiao*

Introduction to Composite Materials Design, Second Edition, *Ever J. Barbero*

Finite Element Analysis of Composite Materials, *Ever J. Barbero*

Strengthening Design of Reinforced Concrete with FRP

Hayder A. Rasheed

CRC Press
Taylor & Francis Group
Boca Raton London New York

CRC Press is an imprint of the
Taylor & Francis Group, an **informa** business

CRC Press
Taylor & Francis Group
6000 Broken Sound Parkway NW, Suite 300
Boca Raton, FL 33487-2742

Version Date: 20140912

International Standard Book Number-13: 978-1-4822-3558-6 (Hardback)

Visit the Taylor & Francis Web site at
http://www.taylorandfrancis.com

and the CRC Press Web site at
http://www.crcpress.com

Dedication

To the memory of my
late father Abdul Sattar Rasheed
To my caring mother Ghania

Contents

Series Preface

Half a century after their commercial introduction, composite materials are of widespread use in many industries. Applications such as aerospace, windmill blades, highway bridge retrofit, and many more require designs that assure safe and reliable operation for 20 years or more. Using composite materials, virtually any property, such as stiffness, strength, thermal conductivity, and fire resistance, can be tailored to the user's needs by selecting the constituent materials, their proportion and geometrical arrangement, and so on. In other words, the engineer is able to design the material concurrently with the structure. Also, modes of failure are much more complex in composites than in classical materials. Such demands for performance, safety, and reliability require that engineers consider a variety of phenomena during the design. Therefore, the aim of the *Composite Materials: Design and Analysis* book series is to bring to the design engineer a collection of works written by experts on every aspect of composite materials that is relevant to their design.

The variety and sophistication of material systems and processing techniques have grown exponentially in response to an ever-increasing number and type of applications. Given the variety of composite materials available as well as their continuous change and improvement, understanding of composite materials is by no means complete. Therefore, this book series serves not only the practicing engineer, but also the researcher and student who are looking to advance the state of the art in understanding material and structural response and developing new engineering tools for modeling and predicting such responses.

Thus, the series is focused on bringing to the public existing and developing knowledge about the material–property relationships, processing–property relationships, and structural response of composite materials and structures. The series scope includes analytical, experimental, and numerical methods that have a clear impact on the design of composite structures.

Preface

The idea of writing this book emerged from a lack of detailed textbook treatments on strengthening design of reinforced concrete members with fiber-reinforced polymer (FRP) despite the large volume of research literature and practical applications that have been contributed since 1987. Even though two attempts to use glass-fiber-reinforced polymer (GFRP) to strengthen concrete members were made in Europe and the United States in the 1950s and 1960s, the technique wasn't successfully applied until 1987, when Ur Meier first strengthened concrete beams with carbon-fiber-reinforced-polymer (CFRP) laminates.

Knowledge in the area of FRP strengthening has matured, culminating with the introduction of specific design guidelines in Canada (ISIS Canada 2001), Europe (FIB Task Group 9.3 2001), and the United States (ACI 440.2R-02), the latter of which was significantly improved after six years in 2008 (ACI 440.2R-08). Today's structural engineer is entitled to a detailed textbook that establishes the art and science of strengthening design of reinforced concrete with FRP beyond the abstract nature of design guidelines. ACI 440.2R-08 provides better guidance than what is typically provided in codes of practice through its "design example" sections. Nevertheless, a textbook that treats the subject of FRP strengthening design with more depth is really needed to introduce it to the civil engineering curriculum.

This textbook has evolved from thorough class notes established to teach a graduate course on "strengthening design of reinforced concrete members with FRP" in spring of 2012 at Kansas State University. The course was widely attended by 18 on-campus senior level, master's level, and doctoral students as well as five distance-education students comprised of practicing engineers pursuing an MS degree. The course included four sets of detailed homework assignments, two term exams, and a research and development project for individuals or teams of two students, depending on the project scope and deliverables, evaluated through project proposals.

Even though the course covered a wide range of topics—from material characterization, flexural strengthening of beams and slabs, shear strengthening of beams, and confinement strengthening of columns, in addition to installation and inspection of FRP as externally bonded (EB) or near-surface-mounted (NSM) composite systems to concrete members—FRP anchorage, FRP strengthening in torsion, and FRP strengthening of prestressed members were left out of the scope of this first book edition. However, it is the intention of the author to add these and other topics to subsequent editions to allow for more selective treatments or more advanced courses to be offered based on this textbook.

The author would like to acknowledge his former graduate student Mr. Augustine F. Wuertz, who helped type a major part of the manuscript while at Kansas State

University. The author would also like to acknowledge Tamara Robinson for editing several chapters of this book as well as the Office of Research and Graduate Programs in the College of Engineering at Kansas State University for providing this editing service. The author would also like to thank Kansas State University for supporting his sabbatical leave during which this book was finalized.

Hayder A. Rasheed
Manhattan, Kansas, USA
Spring 2014

About the Author

Hayder A. Rasheed is professor of civil engineering at Kansas State University. Since 2013, he has held the title of the Thomas and Connie Paulson Outstanding Civil Engineering Faculty Award. Professor Rasheed received his BS degree and his MS degree in civil engineering majoring in structures from the University of Baghdad, Iraq in 1987 and 1990 respectively. In 1996, he received his PhD in civil engineering majoring in structures with a minor in engineering mechanics from the University of Texas at Austin. Professor Rasheed has four years of design and consulting experience in bridges, buildings, water storage facilities, and offshore structures. He spent three-and-a-half years as assistant professor of civil engineering and construction at Bradley University before moving to Kansas State in 2001. Between 2001 and 2013, he moved up the ranks at Kansas State. Since 2010, he has been a fellow of ASCE and a registered professional engineer in the state of Wisconsin since 1998. He authored and co-authored three books and more than 50 refereed journal publications. He developed a number of advanced engineering software packages. He also served as associate editor for the *ASCE Journal of Engineering Mechanics.* He is currently serving on the editorial board of the *International Journal of Structural Stability and Dynamics* and the *Open Journal of Composite Materials.*

About the Author

[illegible]

1 Introduction

1.1 ADVANCEMENTS IN COMPOSITES

Fiber-reinforced polymer (FRP) composites are relatively new compared to conventional construction materials. These composites are manufactured by combining small-diameter fibers with polymeric matrix at a microscopic level to produce a synergistic material. FRP composites have been considered in aerospace applications since the mid-1950s, when they were used in rocket motor casings (Ouellete, Hoa, and Sankar 1986). Because of their light weight and design versatility, they have since entered structural systems in aerospace, automotive, marine, offshore drilling, and civil engineering applications, in addition to sporting goods such as skiing equipment, commercial boats, golf clubs, and tennis rackets (Jones 1975; Gibson 1994; ACI 440R-96 1996).

Typical structural elements made of advanced composites in fighter aircraft include horizontal and vertical stabilizers, flaps, wing skins, and various control surfaces, totaling weight savings of 20% (Gibson 1994). Other important structural elements are helicopter rotor blades. As for the use of advanced composites in commercial aircraft, they enter into the manufacturing of up to 30% of the external surface area (Gibson 1994). However, currently they are only conservatively used in secondary structures in large aircraft.

Advanced composites are used in a variety of additional industries as well. Structural systems constructed of graphite/epoxy composites in space shuttles include cargo bay doors and the solid rocket booster motor case (Gibson 1994). Typical structural elements made of composites in the automotive industry include leaf springs, body panels, and drive shafts (Gibson 1994). Typical pultruded structural shapes are used in lightweight industrial building construction to offer corrosion and electrical/thermal insulation advantages. Another use of advanced composites in civil engineering applications is in the building of lightweight, all-composite, honeycomb-core decks for rapid replacement of short-span bridges (Kalny, Peterman, and Ramirez 2004). Glass FRP (GFRP) reinforcing bars were produced using a pultrusion process created by Marshall Vega Corporation for use with polymer-based concrete in the late 1960s (ACI 440R-96), and these bars continue to improve in their characteristics, such as the addition of helically wound GFRP deformations for enhanced bonding to concrete.

1.2 INFRASTRUCTURE UPGRADE

The transportation infrastructure in the United States and worldwide is aging due to material deterioration and capacity limitations. Since complete rebuilding of such infrastructure requires a huge financial commitment, alternatives of prioritized strengthening and repair need to be implemented. One of the earliest techniques for repair and strengthening of concrete members, dating back to the mid-1970s, involved the use of epoxy-bonded external steel plates (Dussek 1980). However, in the mid-1980s, durability studies revealed that corrosion of external steel plates is a restrictive factor for widespread usage of this technique in external exposure (Van Gemert and Van den Bosch 1985).

A revolutionary advancement in the technique of external strengthening occurred when Meier replaced external steel plates with external carbon FRP (CFRP) plates in 1987. FRP is resistant to corrosion and has high strength-to-weight and high stiffness-to-weight ratios that provide efficient designs and ease of construction. FRP also has excellent fatigue characteristics and is electromagnetically inert. Accordingly, it is a viable replacement to steel in external strengthening applications. Since 1987, research in FRP strengthening techniques has developed an extensive volume of literature proving the effectiveness of the application. The ACI 440 Committee on Fiber Reinforced Polymer Reinforcement has twice reported on state-of-the-art advancements (ACI 440R-96; ACI 440R-07 2007). The same committee has produced two design documents for FRP externally bonded systems for strengthening applications (ACI 440.2R-02; ACI 440.2R-08 2008). The technology has matured to the point that it can be introduced to the structural engineering curriculum through the development of courses and textbooks.

1.3 BEHAVIOR OF STRENGTHENED REINFORCED CONCRETE BEAMS IN FLEXURE

Shallow beams are typically strengthened in flexure by externally bonding FRP plates or sheets on the tension face or soffit of the member, as shown in Figure 1.1. The fibers are oriented along the beam axis in the state-of-the-art application.

FIGURE 1.1 Strengthening the soffit of inverted reinforced concrete beam with CFRP.

Full composite action between the beam and FRP is usually assumed. However, this perfect bond typically depends on the shear stiffness of the interface adhesive (Rasheed and Pervaiz 2002). Most resin adhesives yield excellent bond characteristics with concrete and FRP, leading to perfect composite action. On the other hand, some resin adhesives have low lap shear stiffness, leading to bond slip between FRP and the concrete beam, thus reducing the composite action (Rasheed and Saadatmanesh 2002; Pervaiz and Ehsani 1990). With full composite action, glass FRP (GFRP) and aramid FRP (AFRP) do not increase the initial stiffness of the beam due to their relatively low modulus along the fiber direction. On the other hand, carbon FRP (CFRP) slightly increases the initial stiffness of the beam due to its high modulus along the fiber direction. Accordingly, this application is not used to stiffen the beams; instead, it is used to strengthen the beam due to the high strength of FRP materials available in practice, as shown in Figure 1.2.

Flexural failure modes may be classified as

1. FRP rupture failure after yielding of primary steel reinforcement. This failure mode typically takes place in lightly reinforced, lightly strengthened sections (Arduini, Tommaso, and Nanni [1997], Beam B2).
2. Concrete crushing failure after yielding of primary steel reinforcement. This failure mode typically occurs in moderately reinforced, moderately strengthened sections (Saadatmanesh and Ehsani [1991], Beam A).
3. Cover delamination failure primarily occurring after yielding of steel reinforcement. This failure mode initiates at the FRP curtailment due to stress

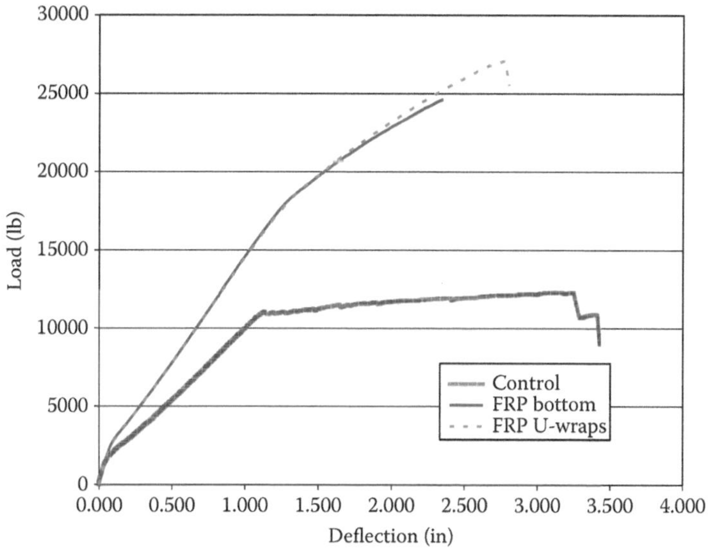

FIGURE 1.2 Response of unstrengthened and CFRP-strengthened identical beams showing limited stiffening compared to strengthening.

concentration at the plate or sheet tip. Once cracking starts at an angle, it changes to a horizontal crack parallel to steel reinforcement at the level of primary steel because the steel stirrups inside the beam arrest the inclined crack. The FRP and the entire concrete cover delaminates (e.g., Arduini, Tommaso, and Nanni [1997], Beam A3 and A4).

4. Plate or sheet debonding along the interface plane due to the intermediate crack mechanism typically after yielding of primary steel reinforcement when the flexural cracks widen. The horizontal crack occurs along the adhesive layer or parallel to it within the concrete cover. This failure mode is especially applicable to beams with end U-wrap anchorage, thus delaying failure in item 3 (e.g., Arduini, Tommaso, and Nanni [1997], Beam B3).
5. Concrete crushing failure for over-reinforced beams or cover delamination failure in beams with short FRP plates prior to primary steel yielding (e.g., Fanning and Kelly [2001], Beam F10).

Shallow beams may also be strengthened with near-surface-mounted (NSM) bars. This strengthening reinforcement is typically made of FRP bars or FRP tape inserted in near-surface cut grooves and then sealed with resin adhesive that fills the groove surrounding the bar or tape, as shown in Figure 1.3 (Rasheed et al. 2010).

FRP in this application behaves similarly to externally bonded FRP plates and sheets. However, failure modes are typically limited to

1. FRP rupture after yielding of primary steel.
2. Concrete crushing after yielding of primary steel.
3. Concrete crushing before yielding of primary steel

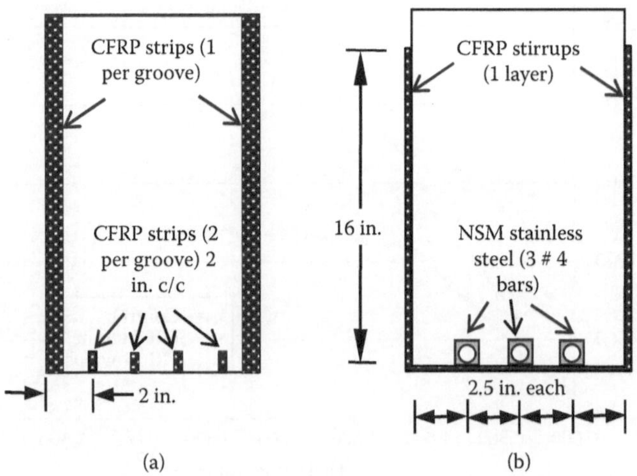

FIGURE 1.3 Strengthening identical beams with (a) CFRP tape and (b) NSM bars.

In other words, cover delamination and FRP debonding are less likely to occur with NSM technology. Highly strengthened sections may suffer from splitting of concrete cover through NSM bars.

The need to use a high ratio of strengthening reinforcement lends itself to combining the two techniques described previously. While no more than three to five layers of FRP sheets or one layer of prefabricated FRP plate may be used as externally bonded reinforcement, combining these external plates or sheets with NSM bars furnishes high strengthening ratios, especially for lightly reinforced sections. This combination also helps the unstrengthened design capacity, since the loss of external strengthening reinforcement still offers higher capacity than the completely unstrengthened section (Rasheed et al. 2013; Traplsi et al. 2013).

1.4 BEHAVIOR OF STRENGTHENED REINFORCED CONCRETE BEAMS IN SHEAR

Concrete beams are typically strengthened in shear by external full wrapping, U-wrapping, or side bonding FRP sheets or fabrics around or along the sides of beams where fibers make 90° or 45° angles with the beam axis along the side profile of the beam, as shown in Figure 1.4. Research and practical applications have shown that this shear strengthening is a highly effective technique. The design of shear-strengthened members with FRP is treated the same way as the design of steel stirrups used as shear reinforcement in beams. The only different design requirement is that the effective FRP strain at failure needs to be identified as opposed to using the yielding strength in the case of steel stirrups.

Failure modes range from FRP fracture for fully wrapped beams to shear debonding for beams with U-wrapped or side-bonded FRP. The effective FRP strain at fracture or debonding is significantly lower than the ultimate tensile strain of FRP due to spots of high stress concentrations, which lead to premature fracture of FRP or peeling off through the concrete near the FRP–concrete interface (Triantafillou 1998). As a result of analyzing the findings of several investigators on this subject, Triantafillou concluded that the effective FRP strain decreases with increasing axial stiffness of the FRP ($\rho_f E_f$). Triantafillou also reported that FRP contribution to shear capacity does not increase beyond $\rho_f E_f$ = 0.4 GPa (58 ksi). ACI 440.2R-08 is used in this text for FRP shear-strengthening design, and

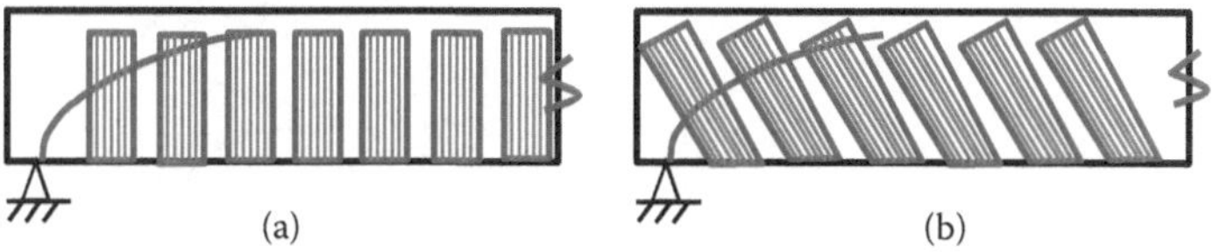

FIGURE 1.4 External shear strengthening of concrete beams: (a) 90° U-wrap sheets and (b) 45° side sheets.

comparisons with experimental results are made to study the conservatism in the code procedure.

1.5 BEHAVIOR OF REINFORCED CONCRETE COLUMNS WRAPPED WITH FRP

Columns in seismic regions must behave in a ductile manner in flexure, shear, and axial directions. One very efficient way to increase this ductility is through the use of FRP wrapping of sheets or fabric such that the main load-carrying fibers are oriented in the hoop direction. This hoop wrapping restricts concrete radial expansion under axial load, leading to columns subjected to confining pressure or a triaxial state of stress, which is known to increase the strength and improve the deformability, as demonstrated in Figure 1.5. Column wrapping with hoop FRP jackets enhances axial behavior. Experimental results and analytical findings confirmed the effectiveness of this technique, especially for seismic upgrade and structural performance of columns subjected to impact (Lam and Teng 2003). Failure modes of wrapped FRP jackets are primarily the FRP fiber rupture at premature levels, debonding, or wrap-unwinding at some point during the loading (Hart 2008), as shown in Figure 1.6. Accordingly, an effective axial strain needs to be established experimentally, at which point circumferential strain is critical (ACI 440.2R-08).

Another application similar to column retrofitting but used for new construction is the utilization of concrete-filled FRP tubes. In this application, flexural performance, shear capacity, compressive strength, and strain performance are enhanced due to the significant stiffness of the tube in the longitudinal (axial) direction, contributing to the composite action of the section. Extra stiffness in the hoop direction contributes to the confinement and additional shear capacity of the concrete-filled FRP tube.

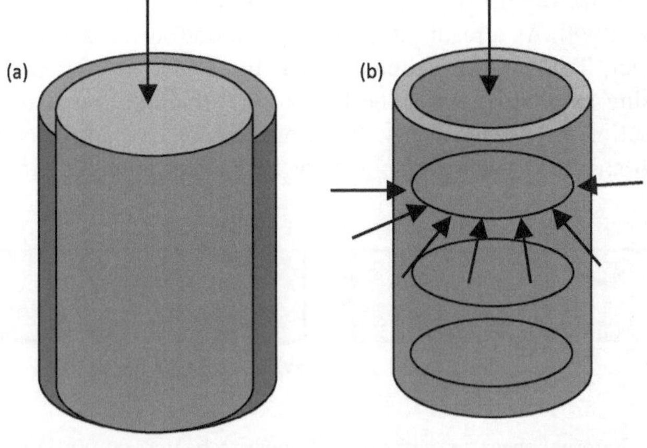

FIGURE 1.5 Confinement in FRP-wrapped columns: (a) unconfined column and (b) concrete column confined with wrapped FRP.

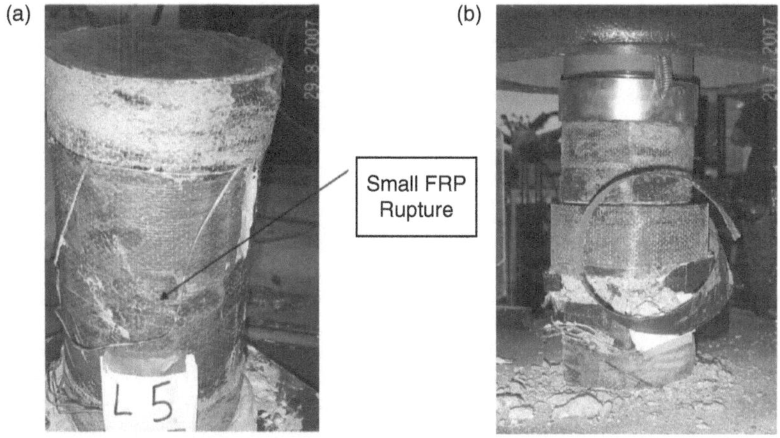

FIGURE 1.6 Failure modes in columns wrapped with CFRP: (a) FRP rupture and (b) FRP debonding and unwinding. (Photo courtesy of Hart [2008].)

REFERENCES

ACI 440R-96. 1996. Report on fiber-reinforced plastic (FRP) reinforcement for concrete structures. ACI Committee 440, Farmington Hills, MI.

ACI 440R-07. 2007. Report on fiber-reinforced polymer (FRP) reinforcement for concrete structures. ACI Committee 440, Farmington Hills, MI.

ACI 440.2R-02. 2002. Guide for the design and construction of externally bonded FRP systems for strengthening concrete structures. ACI Committee 440, Farmington Hills, MI.

ACI 440.2R-08. 2008. Guide for the design and construction of externally bonded FRP systems for strengthening concrete structures. ACI Committee 440, Farmington Hills, MI.

Arduini, M., A. Tommaso, and A. Nanni. 1997. Brittle failure in FRP plate and sheet bonded beams. *ACI Structural Journal* 94 (4): 363–70.

Dussek, I. J. 1980. Strengthening of bridge beams and similar structures by means of epoxy-resin-bonded external reinforcement. Transportation Research Record 785, Transportation Research Board, 21–24.

Fanning, P., and O. Kelly. 2001. Ultimate response of RC beams strengthened with CFRP plates. *ASCE Journal of Composites in Construction* 5 (2): 122–27.

Gibson, R. F. 1994. *Principles of composite material mechanics*. New York: McGraw Hill.

Hart, S. D. 2008. Performance of confined concrete columns under simulated life cycles. PhD diss., Kansas State Univ., Manhattan, KS.

Jones, R. M. 1975. *Mechanics of composite materials*. New York: Hemisphere Publishing.

Kalny, O., R. J. Peterman, and G. Ramirez. 2004. Performance evaluation of repair technique for damaged fiber-reinforced polymer honeycomb bridge deck panels. *ASCE Journal of Bridge Engineering* 9 (1): 75–86.

Lam, L., and J. G. Teng. 2003. Design-oriented stress–strain model for FRP-confined concrete. *Construction and Building Materials* 17 (6-7): 471–89.

Meier, U. 1987. Bridge repair with high performance composite materials. *Material und Technik* 15 (4): 125–28.

Ouellette, S. V., T. S. Hoa, and T. S. Sankar. 1986. Buckling of composite cylinders under external pressure. *Polymer Composites* 7 (5): 363–74.

Rasheed, H. A., R. R. Harrison, R. J. Peterman, and T. Alkhrdaji. 2010. Ductile strengthening using externally bonded and near surface mounted composite systems. *Composite Structures* 92 (10): 2379–90.

Rasheed, H. A., and S. Pervaiz. 2002. Bond slip analysis of FRP-strengthened beams. *ASCE Journal of Engineering Mechanics* 128 (1): 78–86.

Rasheed, H. A., A. Wuertz, A. Traplsi, H. G. Melhem, and T. Alkhrdaji. 2013. Externally bonded GFRP and NSM steel bars for improved strengthening of rectangular concrete beams. In *Advanced materials and sensors toward smart concrete bridges: Concept, performance, evaluation, and repair*. SP-298. Farmington Hills, MI: American Concrete Institute.

Saadatmanesh H., and M. Ehsani. 1990. Fiber composite plates can strengthen beams. *Concrete International* 12 (3): 65–71.

Saadatmanesh H., and M. Ehsani. 1991. RC beams strengthened with GFRP plate, Part 1: Experimental study. *ASCE Journal of Structural Engineering* 117 (11): 3417–33.

Traplsi, A., A. Wuertz, H. Rasheed, and T. Alkhrdaji. 2013. Externally bonded GFRP and NSM steel bars for enhanced strengthening of concrete T-beams. In *TRB 92nd Annual Meeting Compendium of Papers*. Paper 13-3120. Washington, DC: Transportation Research Board.

Triantafillou, T. C. 1998. Shear strengthening of reinforced concrete beams using epoxy-bonded FRP composites. *ACI Structural Journal* 95 (2): 107–15.

Van Gemert, D. A., and M. C. J. Van den Bosch. 1985. Repair and strengthening of reinforced concrete structures by means of epoxy bonded steel plates. *International Conference on Deterioration*, Bahrain, 181–92.

2 Background Knowledge

2.1 OVERVIEW

Before starting the discussion on FRP strengthening, it is important to refresh the basics of concrete design of sections and members upon which the FRP strengthening design equations are built. Consequently, this chapter revisits the following four background topics:

1. Flexural design of RC sections
2. Shear design of RC beams
3. Internal reinforcement to confine RC columns
4. Service load calculations in beams

The inclusion of these four topics was based on the fact that they represent the primary structural strengthening subjects addressed in this book.

2.2 FLEXURAL DESIGN OF RC SECTIONS

The fundamental principles of flexural design are strain compatibility, force, and moment equilibrium, as well as material constitutive (stress–strain) relationships.

2.2.1 Strain Compatibility

Shallow-beam theory accurately assumes that plane sections before bending remain plane after bending, directly translating into linear strain distribution across the beam section until ultimate failure of that section. Typical reinforced concrete sections fail in flexure by concrete crushing when concrete compressive strain reaches around 0.003 after the yielding of primary tensile reinforcement. The value of 0.003 is selected by the American Concrete Institute code to mark the attainment of concrete crushing (ACI 318-11). This basic failure mode is a ductile one, since the member shows high deformability prior to reaching ultimate capacity. Conversely, beams may fail in flexure by concrete crushing prior to the yielding of primary steel, which is an undesirable brittle failure that is not allowed by ACI 318-11 code for beams, while the latter failure mode is allowed for other structural members by reducing the ϕ factor significantly (i.e., increasing the margin of strength), as shown in Figure 2.1 for sections with grade 60 reinforcement.

ACI 318-11 section 10.3.5 states that for members with factored axial compressive load less than $0.1f_c'A_g$ (beams), the strain at the extreme level of reinforcement (ε_t) at nominal strength shall not be less than 0.004, yielding a ductile failure of the beam,

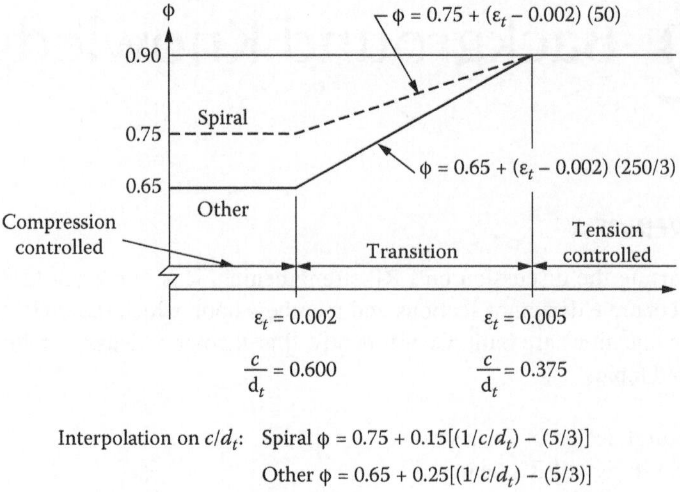

FIGURE 2.1 Change of ϕ factor with net tensile strain in extreme steel bars (ε_t) or neutral-axis depth ratio (c/d_t) for Grade 60 reinforcement. (Courtesy of ACI 318-11.)

as seen in Figure 2.2b. The strain profile at steel yielding and ultimate capacity are illustrated in Figure 2.2. At a level of tensile steel strain ($\varepsilon_t = 0.004$) when compressive extreme concrete fiber strain reaches concrete crushing ($\varepsilon_{cu} = 0.003$), the value of $c/dt = 0.429$ results from similar triangles (Figure 2.2b), which corresponds to a factor $\phi = 0.817$ for members other than those with spiral steel (Figure 2.1).

2.2.2 Force Equilibrium

To determine the location of the neutral axis in beams, force equilibrium needs to be satisfied as follows:

Analysis problem:

$$C = T \tag{2.1}$$

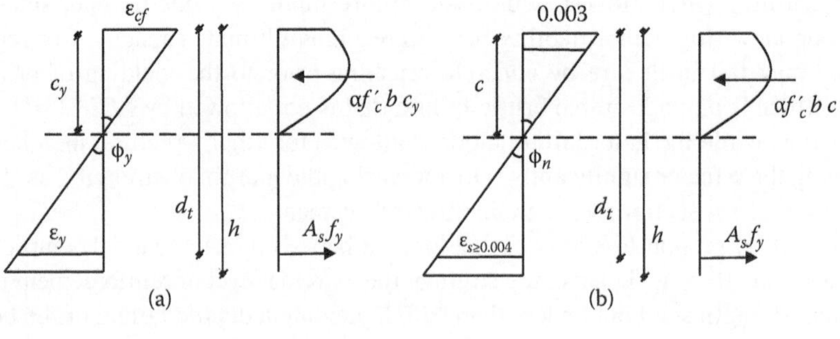

FIGURE 2.2 Strain and force profile at (a) first steel yielding and (b) ultimate capacity.

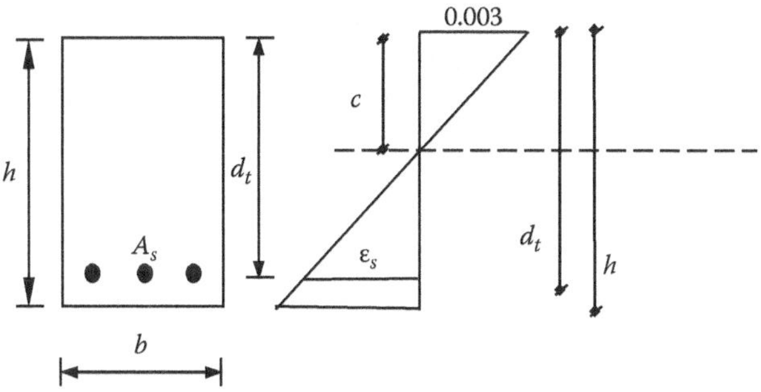

FIGURE 2.3 Typical rectangular section and strain profile at section failure.

$$0.85 f_c' b \beta_1 c = A_s f_y \tag{2.2}$$

$$c = \frac{A_s f_y}{0.85 f_c' b \beta_1} \text{ (See Figure 2.3.)} \tag{2.3}$$

If $\frac{c}{d_t} > 0.429 \rightarrow$ This is a Brittle Failure N. G.

Design problem:

$$C = T \tag{2.4}$$

$$0.85 f_c' b \beta_1 c = A_s f_y \tag{2.5}$$

$$c = \frac{A_s f_y}{0.85 f_c' b \beta_1} \text{ (See Figure 2.3.)} \tag{2.6}$$

The value c is unknown and is to be substituted into the moment equation to solve for the area of steel, A_s, (while $\frac{c}{d_t} < 0.429$).

2.2.3 Moment Equilibrium

To determine the moment-carrying capacity of the beam section in analysis problems or the steel area in design problems, the moment-equilibrium equation is involved:

$$M_u = \phi A_s f_y \left(d - \frac{a}{2} \right) \tag{2.7}$$

where

$$a = \beta_1 c$$

$$M_n = \frac{M_u}{\phi}$$

$$M_n = \rho b d^2 f_y \left(1 - \frac{a}{2d}\right) \tag{2.8}$$

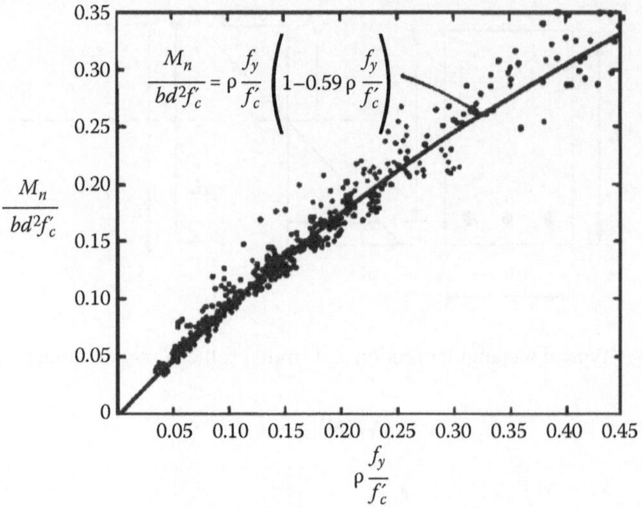

FIGURE 2.4 Tests from 364 beams governed by tension. (Courtesy of Portland Cement Association, 2013.)

Substituting $a = \beta_1 c$ from the force equilibrium, Equation (2.3),

$$\frac{M_n}{bd^2 f'_c} = \rho\frac{f_y}{f'_c}\left(1 - \frac{A_s f_y}{2*0.85bdf'_c}\right)$$

$$\frac{M_n}{bd^2 f'_c} = \rho\frac{f_y}{f'_c}\left(1 - 0.59\rho\frac{f_y}{f'_c}\right) \quad \text{(See Figure 2.4)} \tag{2.9}$$

$$R = \omega(1 - 0.59\omega) \tag{2.10}$$

where

$$\omega = \rho\frac{f_y}{f'_c}, \quad R = \frac{M_n}{bd^2 f'_c}$$

$$R = \omega - 0.59\omega^2 \quad \text{(Analysis Equation)} \tag{2.11}$$

$$0.59\omega^2 - \omega + R = 0$$

$$\omega = \frac{1 - \sqrt{1 - 2.36R}}{1.18} \quad \text{(Design Equation)} \tag{2.12}$$

2.2.4 Constitutive Relationships

Concrete is strong in compression and weak in tension because of structural cracking at a relatively low level of stress. Therefore, concrete needs to be reinforced with steel in tension to bridge cracks and carry tension. Concrete and steel are thermally compatible, making them ideal for a composite material. Reinforcing steel has a coefficient of thermal expansion of 6.5 × 10^{-6}/°F (11.7 × 10^{-6}/°C), while concrete has a coefficient of thermal expansion of 5.5 × 10^{-6}/°F (9.9 × 10^{-6}/°C) (Beer and Johnston 1992). The constitutive behavior of reinforced concrete constituent materials is described in the following sections.

2.2.4.1 Behavior of Concrete in Compression

The stress–strain response of concrete in compression is nearly linear at the beginning of loading up to approximately $0.7f_c'$. Beyond that stress, the response becomes highly nonlinear up to failure. One of the simplest models to effectively capture the stress–strain response of concrete in compression is that of Hognestad's parabola (Hognestad 1951), shown in Figure 2.5.

$$\sigma_c = f_c'\left[2\frac{\varepsilon_c}{\varepsilon_c'} - \left(\frac{\varepsilon_c}{\varepsilon_c'}\right)^2\right] \quad 0 < \varepsilon_c < \varepsilon_{cu} = 0.003 \tag{2.13}$$

where ε_c' is the strain corresponding to f_c', typically equal to 0.002 for normal-strength concrete or, more accurately, $\varepsilon_c' = 1.71\frac{f_c'}{E_c}$ (MacGregor 1992). The variable f_c' is the 28-day compressive strength of standard 6 × 12 in. (150 × 300 mm) cylinders, and ε_{cu} is the limit of useful compression strain of concrete before crushing, identified as 0.003 by ACI 318-11, as demonstrated in Figure 2.6.

The initial tangent modulus from Hognestad's parabola is

$$\left.\frac{d\sigma}{d\varepsilon_c}\right|_{\varepsilon_c=0} = E_{ci} = f_c'\left[\frac{2}{\varepsilon_c'} - \frac{2\varepsilon_c}{\varepsilon_c'^2}\right] \cong 1000f_c' \tag{2.14}$$

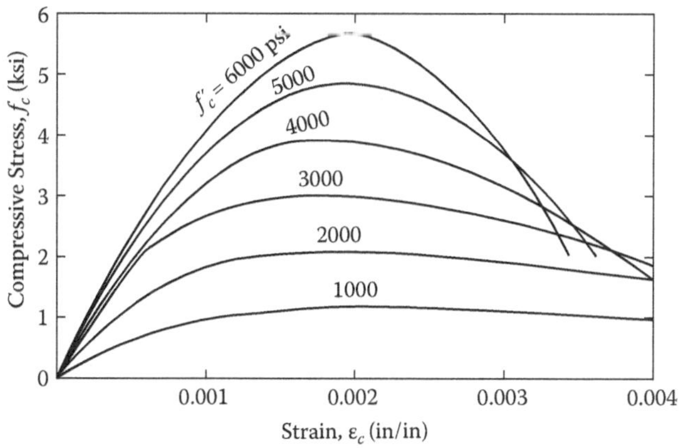

FIGURE 2.5 Typical stress-strain curves for concrete of various strengths. (Courtesy of Portland Cement Association [2013].)

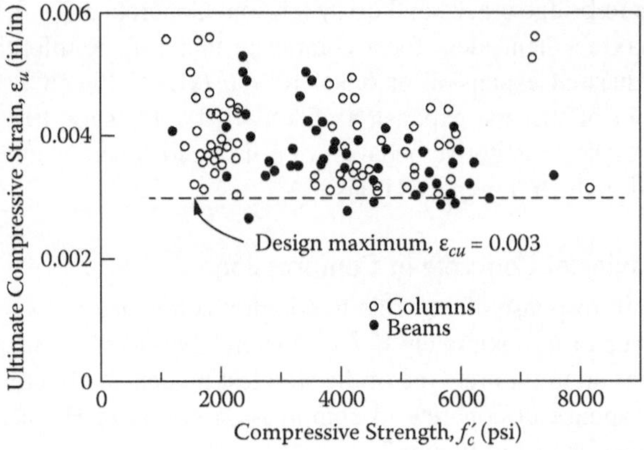

FIGURE 2.6 Tests of reinforced concrete members showing selection of the design maximum compressive strain. (Courtesy of Portland Cement Association [2013].)

However, Young's modulus for concrete is taken as the secant value at $0.4f_c'$. This is given by the following ACI 318-11 equation:

$$E_c = 57\sqrt{f_c'} \text{ (US customary units, } E_c \text{ in ksi, } f_c' \text{ in psi)} \tag{2.15a}$$

$$E_c = 4700\sqrt{f_c'} \text{ (SI units, } E_c \text{ and } f_c' \text{ in MPa)} \tag{2.15b}$$

For concrete between 3000–5000 psi, $E_{ci} \cong 3000 - 5000$ ksi, $E_c = 3122 - 4031$ ksi, so the two moduli, are comparable in this range of f_c' values. In flexural calculations, ACI allows the usage of Whitney's rectangular stress block to replace Hognestad's parabola. At the concrete crushing failure point, ACI assigns the block a height factor of $\gamma = 0.85$ based on tests of columns (Hognestad 1951) and the effect of sustained load on the strength of concrete (Rüsch 1960). Also, the effective depth factor is

$$\beta_1 = 0.85 \quad \text{if} \quad f_c' \le 4000 \text{ psi} \quad \text{or} \quad f_c' \le 30 \text{ MPa} \tag{2.16}$$

$$\begin{aligned} \beta_1 &= 1.05 - 0.05\frac{f_c'}{1000} \quad \text{if} \quad 4000 \text{ psi} < f_c' < 8000 \text{ psi} \\ \beta_1 &= 1.09 - 0.008 f_c' \quad \text{if} \quad 30 \text{ MPa} < f_c' < 55 \text{ MPa} \end{aligned} \tag{2.17}$$

$$\beta_1 = 0.65 \quad \text{if} \quad f_c' \ge 8000 \text{ psi} \quad \text{or} \quad f_c' \ge 55 \text{ MPa} \tag{2.18}$$

This was selected by comparison with experimental data points (Kaar, Hanson, and Capell 1978), as seen in Figure 2.7.

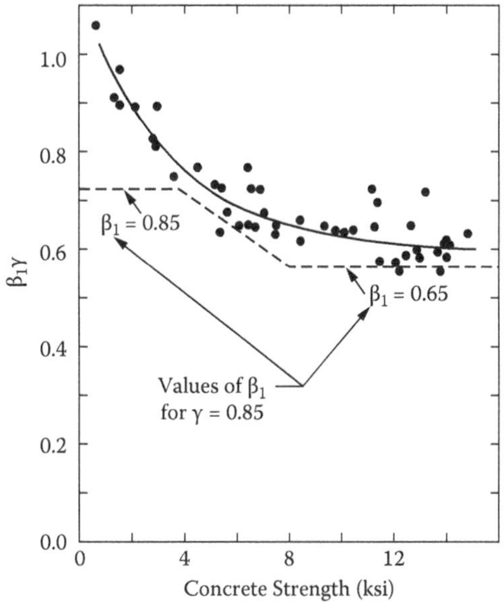

FIGURE 2.7 Selection of β_1 variation with concrete strength. (Courtesy of ACI SP-55.)

Total compressive force C may be expressed as follows:

$$C = \alpha f_c' bc \tag{2.19}$$

To derive the expression of α based on Hognestad's parabola, the concept of replacing the area under the parabola with an equivalent area of a rectangular block with a height of $\alpha f_c'$ is introduced (Park and Paulay 1975):

$$\alpha f_c' \varepsilon_{cf} = \int_0^{\varepsilon_{cf}} \sigma_c \, d\varepsilon_c = f_c' \int_0^{\varepsilon_{cf}} \left(2\frac{\varepsilon_c}{\varepsilon_c'} - \left(\frac{\varepsilon_c}{\varepsilon_c'} \right)^2 \right) d\varepsilon_c$$

$$\alpha \varepsilon_{cf} = \left[\frac{\varepsilon_c^2}{\varepsilon_c'} - \frac{\varepsilon_c^3}{3\varepsilon_c'^2} \right]_0^{\varepsilon_{cf}} = \frac{\varepsilon_{cf}^2}{\varepsilon_c'} - \frac{\varepsilon_{cf}^3}{3\varepsilon_c'^2} \Rightarrow \alpha = \frac{\varepsilon_{cf}}{\varepsilon_c'} - \frac{1}{3}\left(\frac{\varepsilon_{cf}}{\varepsilon_c'} \right)^2 \tag{2.20}$$

For $\varepsilon_{cf} = 0.003$ and $\varepsilon_c' = 1.71 f_c'/E_c$, values of the factor ($\alpha$) are shown in Table 2.1. The equivalent rectangular block, according to ACI 318-11, has the following expression, as demonstrated in Figure 2.8 and Table 2.1:

$$C = \gamma f_c' ba = \gamma f_c' b\beta_1 c \Rightarrow \alpha = \gamma \beta_1 \Rightarrow \gamma = \frac{\alpha}{\beta_1} \tag{2.21}$$

TABLE 2.1
Variations of Factors α and γ with Concrete Strength

f'_c (psi)	ε'_c [a]	α	γ
3000	0.00164	0.715	0.84
4000	0.0019	0.748	0.88
5000	0.00212	0.748	0.94
6000	0.00232	0.736	0.98
7000	0.00251	0.719	1.03
8000	0.00268	0.701	1.08

[a] Equation from MacGregor (1992).

It is evident from Table 2.1 that values of (γ) noticeably exceed the ACI value of 0.85 as the concrete strength increases beyond 4000 psi. This is attributed to higher (α) values obtained from Equation (2.20) compared to those selected by Kaar, Hanson, and Capell (1978) for higher f'_c values as seen in Table 2.2.

Sources of inconsistency in α and γ results between Tables 2.1 and 2.2 are attributed to the fact that α and β_1 in Table 2.2 are selected by Kaar, Hanson, and Capell (1978) to be a lower bound of all the scattered experimental points, whereas it is analytically computed in Table 2.1.

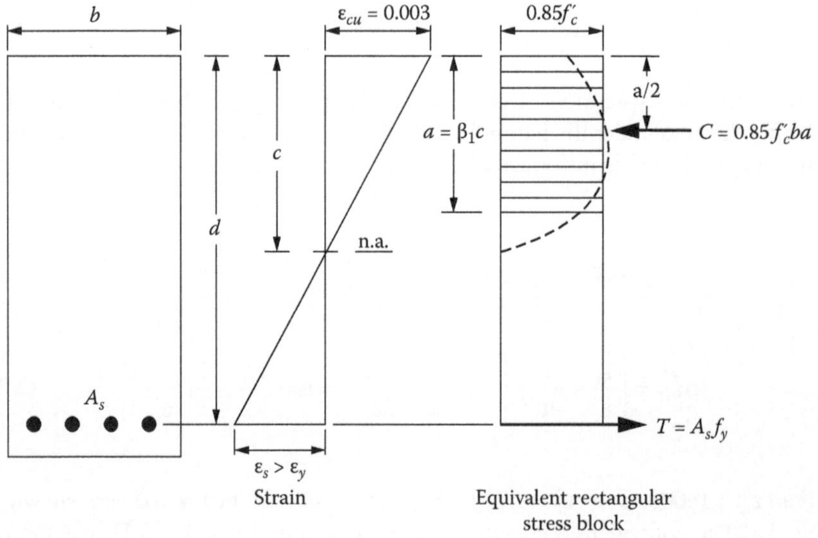

FIGURE 2.8 ACI equivalent rectangular stress block. (Courtesy of Portland Cement Association [2013].)

TABLE 2.2
Selection of Factors α and β_1

	f_c', psi				
	≤4000	**5000**	**6000**	**7000**	**≥8000**
α	0.72	0.68	0.64	0.60	0.56
β	0.425	0.400	0.375	0.350	0.325
$\beta_1 = 2\beta$	0.85	0.80	0.75	0.70	0.65
$\gamma = \alpha/\beta_1$	0.85	0.85	0.85	0.86	0.86

Source: Kaar, Hanson, and Capell (1978).

2.2.4.2 Behavior of Concrete in Tension

Concrete experiences very little hardening or nonlinear plasticity, if any, prior to cracking or fracture. The ultimate strength of concrete in tension is relatively low, and cracked member behavior is significantly different, which is why the prediction of cracking strength is critical. Three distinct tests estimate concrete tensile strength: direct tension test, split-cylinder test, and flexure test. In direct tension test, stress concentration at the grips and load axis misalignment yield lower strength. The split-cylinder test uses a 6×12 in. (150×300 mm) standard cylinder on its side subjected to vertical compression generating splitting tensile stresses of $\frac{2P}{\pi dL}$. More reasonable estimates of tensile strength are generated by the split-cylinder test. A flexure test is the most widely used test to measure the modulus of rupture (f_r). This test assumes that concrete is elastic at fracture, and the bending stress is known to be localized at the tension face. Accordingly, results are expected to slightly overestimate tensile strength. According to ACI 318-11, the modulus of rupture (f_r) is

$$f_r = 7.5\lambda\sqrt{f_c'} \quad \text{in psi} \tag{2.22}$$

$$f_r = 0.62\lambda\sqrt{f_c'} \quad \text{in MPa} \tag{2.23}$$

For normal weight concrete, $\lambda = 1.0$ (Section 8.6.1, ACI 318-11)
For sand-lightweight concrete, $\lambda = 0.85$
For all lightweight concrete, $\lambda = 0.75$
If f_{ct} is given for lightweight concrete, $\lambda = \frac{f_{ct}}{6.7\sqrt{f_c'}} \leq 1.0$

The value of λ can be determined from ACI 318-11 based on two alternative approaches presented in commentary R8.6.1. Typical values for direct tensile strength (f_t') for normal weight concrete are $3-5\sqrt{f_c'}$ in psi and for light weight concrete are

$2-3\sqrt{f_c'}$ in psi (Nilson 1997). On the other hand, typical values for split-cylinder strength (f_{ct}) for normal weight concrete are $6-8\sqrt{f_c'}$ in psi and for light weight concrete are $4-6\sqrt{f_c'}$ in psi. Furthermore, typical values for modulus of rupture (f_r) for normal weight concrete are $8-12\sqrt{f_c'}$ in psi and for light weight concrete are $6-8\sqrt{f_c'}$ in psi. The tensile contribution of uncracked concrete is ignored at the level of ultimate capacity because it is negligible at that stage.

2.2.4.3 Behavior of Reinforcing Steel

Steel reinforcement bars behave similarly in tension and compression. Even though mild steel has some strain hardening prior to final fracture, the ACI 318 code assumes a flat plateau after steel yielding, thus conservatively ignoring this strain hardening effect, as shown in Figure 2.9. On the other hand, higher strength steels have non-linear strain hardening behavior after steel yielding, as demonstrated in Figure 2.10. However, typical design computations still model steel as elastic–perfectly plastic to conservatively simplify the calculations.

The primary parameters that define the idealized stress–strain model of reinforcing steel are

- E_s: The Young's modulus of elasticity for steel is known to be approximately equal to 29,000 ksi = 200,000 MPa.
- f_y: The yield strength of steel, which varies depending on the composition of steel alloy, ranges between 40–100 ksi (276–690 MPa).

For high-strength steel, ACI 318-11 code specifies f_y as the stress at $\varepsilon_s = 0.0035$ ($f_y > 60$ ksi, 414 MPa), as shown in Figure 2.10.

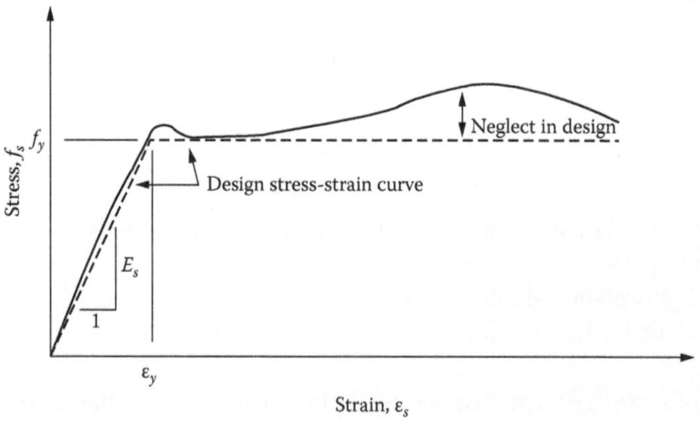

FIGURE 2.9 Actual and idealized stress-strain curve of reinforcing steel. (Courtesy of Portland Cement Association [2013].)

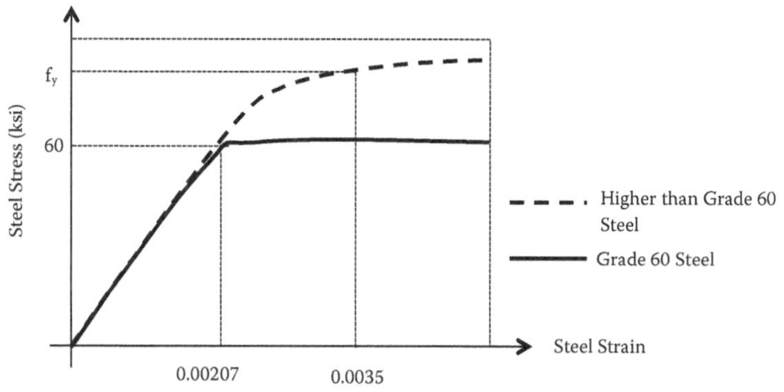

FIGURE 2.10 Typical stress–strain response of reinforcing steel.

Example 2.1: Analysis

For the reinforced concrete beam section shown in Figure 2.11, determine the ultimate moment capacity by neglecting the compression reinforcement.

$$f_c' = 4 \text{ ksi}$$
$$f_y = 60 \text{ ksi}$$

Solution:

First, calculate the effective depth and area of tensile reinforcement using Table A-1 of Appendix A.

$$d = d_t = 18 - 1.5 - \frac{4}{8} - \frac{1}{2} \times 0.75 = 15.625 \text{ in.}$$

$$A_s = 4 \times 0.44 = 1.76 \text{ in.}^2$$

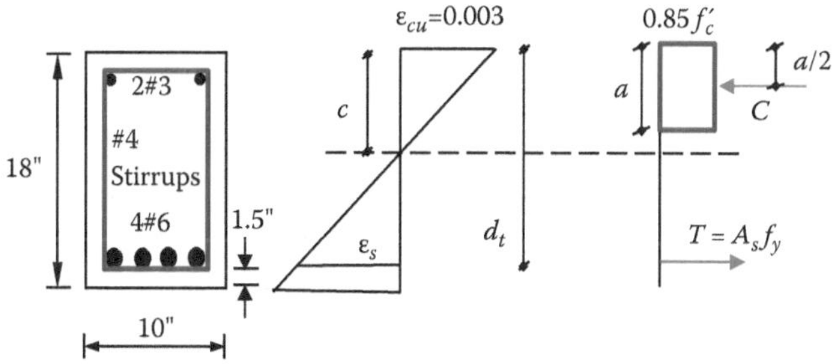

FIGURE 2.11 Cross-section details, strain, and force profile for Example 2.1.

Assume the tensile steel has yielded at ultimate capacity

$$a = \frac{A_s f_y}{0.85 f_c' b} = \frac{1.76 \times 60}{0.85 \times 4 \times 10} = 3.11\,\text{in.}$$

$$\Rightarrow \beta_1 = 0.85 \Rightarrow c = \frac{a}{\beta_1} = 3.65\ \text{in.}$$

$$\frac{c}{d_t} = \frac{3.65}{15.625} = 0.234 < 0.375 \Rightarrow \text{ tension controlled failure} \Rightarrow \phi = 0.9$$

Or determine the actual strain in tension steel using strain compatibility:

$$\frac{\varepsilon_s}{d_t - c} = \frac{0.003}{c} \Rightarrow \varepsilon_s = 0.00984 > 0.005 \Rightarrow \phi = 0.9$$

$$\varepsilon_s = 0.00984 > 0.004 \quad \text{O.K.}$$

Computing the section moment capacity:

$$M_u = \phi A_s f_y \left(d - \frac{a}{2}\right) = 0.9 \times 1.76 \times 60 \times \left(15.625 - \frac{3.11}{2}\right) = 1337.2\ \text{k-in.} = 111.43\ \text{k-ft}$$

$$A_{s\min} = \frac{3\sqrt{f_c'}}{f_y} b_w d \geq \frac{200}{f_y} b_w d$$

$$\frac{3\sqrt{f_c'}}{f_y} = 0.00316 < \frac{200}{f_y} = 0.00333 \quad \text{the latter controls}$$

$$A_{s\min} = 0.00333 \times 10 \times 15.625 = 0.52\ \text{in.}^2 < 1.76\ \text{in.}^2 \quad \text{O.K.}$$

Example 2.2: Design

For the reinforced concrete beam section shown in Figure 2.12, design the doubly reinforced section to resist a moment capacity of $M_u = 220$ k-ft, knowing that the primary steel is composed of #7 bars and the compression steel is composed of #4 bars. Assume the shear stirrup size is #4 bar.

$$f_c' = 4\ \text{ksi}$$
$$f_y = 60\ \text{ksi}$$

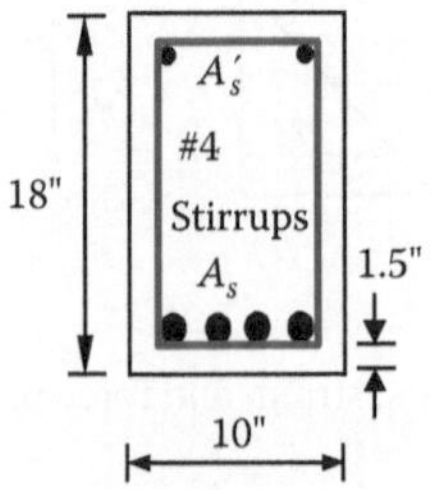

FIGURE 2.12 Cross-section details for Example 2.2.

Solution:

$$d = d_t = 18'' - 1.5'' - \frac{4}{8} - \frac{1}{2} \times 0.875 = 15.563''$$

$$d' = 1.5'' + \frac{4}{8} + \frac{1}{2} \times \frac{4}{8} = 2.25''$$

Assume $\phi = 0.9 \Rightarrow \varepsilon_t = 0.005$

$$\frac{0.003}{c} = \frac{0.005}{d_t - c} \Rightarrow \frac{d_t}{c} - 1 = 1.67$$

$$\frac{d_t}{c} = 2.67 \Rightarrow \frac{c}{d_t} = 0.375 \text{ (see Figure 2.1)}$$

$$c = 0.375 d_t = 0.375 \times 15.563 = 5.84 \text{ in.}$$

$$A_s = \frac{0.85 f_c' b \beta_1 c}{f_y} = 2.813 \text{ in.}^2$$

$$M_{ut} = \phi A_s f_y \left(d - \frac{a}{2} \right) = 1987 \text{ k-in.} = 165.58 \text{ k-ft} < 220 \text{ k-ft}$$

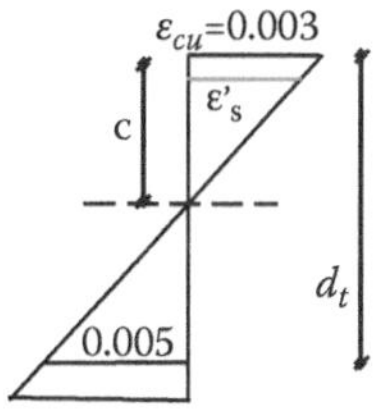

Thus, enlarge section or use compression reinforcement.

Another solution approach is possible using Table 2.3 from Portland Cement Association (2013). R_{nt} from Table 2.3 is 911 psi. This R_{nt} can be simply calculated as follows:

$$R_{nt} = \frac{M_{nt}}{bd^2} = \frac{M_{ut}}{\phi b d_t^2} = \frac{165.58 \times 12}{0.9 \times 10 \times 15.563^2} = 0.9115 \text{ ksi} = 911.5 \text{ psi}$$

$$R_n = \frac{M_n}{bd_t^2} = \frac{M_u}{\phi b d_t^2} = \frac{220 \times 12 \times 1000}{0.9 \times 10 \times 15.563^2} = 1211.1 > 911 \text{psi}$$

TABLE 2.3
Design Parameters at Steel Strain of 0.005 for Tension-Controlled Sections

		$f_c' = 3{,}000$	$f_c' = 4{,}000$	$f_c' = 5{,}000$	$f_c' = 6{,}000$	$f_c' = 8{,}000$	$f_c' = 10{,}000$
		$\beta_1 = 0.85$	$\beta_1 = 0.85$	$\beta_1 = 0.80$	$\beta_1 = 0.75$	$\beta_1 = 0.65$	$\beta_1 = 0.65$
	R_{nt}	683	911	1084	1233	1455	1819
	ϕR_{nt}	615	820	975	1109	1310	1637
	ω_t	0.2709	0.2709	0.2550	0.2391	0.2072	0.2072
ρ_t	Grade 40	0.02032	0.02709	0.03187	0.03586	0.04144	0.05180
	Grade 60	0.01355	0.01806	0.02125	0.02391	0.02762	0.03453
	Grade 75	0.01084	0.01445	0.01700	0.01912	0.02210	0.02762

Source: Courtesy of Portland Cement Association (2013).

Design for doubly reinforced section.

$$M_{nt} = R_{nt}bd_t^2 = 911\times10\times15.563^2 = 2,206,505.5 \text{ lb-in.} = 2206.5 \text{ k-in.} = 183.88 \text{ k-ft}$$

$$M_n' = M_n - M_{nt} = \frac{M_u}{\phi} - M_{nt} = \frac{220}{0.9} - 183.88 = 60.56 \text{ k-ft}$$

Strain in compression steel:

$$\frac{\varepsilon_s'}{c-d'} = \frac{0.003}{c} \Rightarrow \varepsilon_s' = \frac{5.84-2.25}{5.84}\times0.003 = 0.001844 < \frac{60}{29,000} = 0.00207$$

So compression steel does not yield

$$f_s' = E_s\varepsilon_s' = 29,000\times0.001844 = 53.48 \text{ ksi}$$

$$M_u' = \phi A_s' f_s'(d-d')$$

$$60.56\times12 = A_s'\times53.48\times(15.563-2.25) \Rightarrow A_s' = 1.021 \text{ in.}^2$$

$$A_s = A_{st} + A_s'\frac{f_s'}{f_y} = 2.813 + 1.021\times\frac{53.48}{60} = 3.723 \text{ in.}^2$$

or use Table 2.3 to determine ρ_{st}

$$\rho_{st} = 0.01806$$

$$A_s = \rho_{st}bd + A_s'\frac{f_s'}{f_y} = 3.721 \text{ in.}^2$$

#7 bars (0.6 in.^{2}) $\Rightarrow \frac{3.721}{0.6} = 6.2 \Rightarrow$ Use seven #7 bars for tension

$b = 1.5\times2 + 0.5\times2 + 7\times0.875 + 6\times1 = 16.125\text{"} > 10\text{"} \Rightarrow$ Use two layers (ACI Section 7.6)

$b = 1.5\times2 + 0.5\times2 + 4\times0.875 + 3\times1 = 12 > 10\text{"} \Rightarrow$ Use three layers (ACI Section 7.6)

#7 bars (0.6 in.^{2}) $\Rightarrow \frac{1.021}{0.6} = 1.7 \Rightarrow$ Use two #7 bars for compression (better fits the width)

It is evident that the bar arrangement in Figure 2.13 has caused a slightly higher bar area, smaller effective steel depth (d), and larger compression steel depth (d'), all of which may or may not furnish the moment capacity specified in Example 2.2. Accordingly, the reader may check that by solving Problem 2.3 at the end of the chapter.

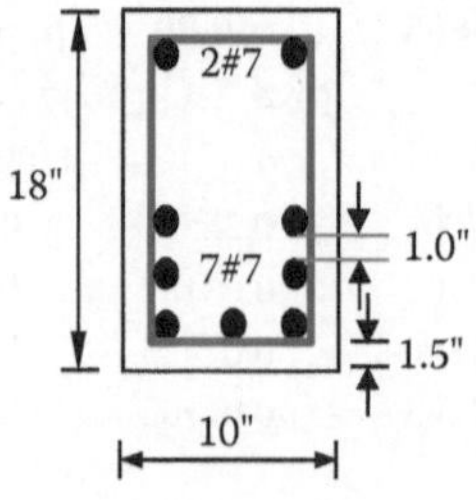

FIGURE 2.13 Cross-section design for Example 2.2.

2.3 SHEAR DESIGN OF RC BEAMS

In addition to flexural failure of beams, which is typically ductile in nature and provides warning signs of large deflections as well as increased and widened flexural cracks with continuous yielding of primary steel, beams may fail in shear or diagonal tension, which is a sudden brittle failure posing more threat than flexural failure. Furthermore, accurate prediction of shear failure is difficult to achieve because the mechanisms involved are not all completely understood. Similar to strategies to protect against flexural failure, concrete beams are typically reinforced with shear reinforcement (stirrups), which are uniformly distributed along the beam profile in a vertical or inclined orientation to provide bridging of diagonal tension cracks and control or delay shear failure, such that flexural failure takes place first. Accordingly, ACI 318-11 specifies a lower strength reduction factor for shear ($\phi = 0.75$) compared to tension-controlled flexural failure ($\phi = 0.9$). Reduced shear resistance must exceed factored shear demand:

$$\phi V_n \geq V_u \tag{2.24}$$

but

$$V_n = V_c + V_s \tag{2.25}$$

thus

$$V_u \leq \phi V_c + \phi V_s \tag{2.26}$$

where in the case of shear and flexure only

$\phi = 0.75$

$$V_c = 2\lambda\sqrt{f_c'}b_w d \quad \text{(see Figure 2.14)} \tag{2.27}$$

$$V_c = \left(1.9\lambda\sqrt{f_c'} + 2500\rho_w \frac{V_u d}{M_u}\right) b_w d \leq 3.5\lambda\sqrt{f_c'}\mathrm{b_w d} \quad \text{and} \quad \frac{V_u d}{M_u} \leq 1.0 \tag{2.28}$$

while in the case of shear, flexure, and axial compression.

$$V_c = 2\left(1 + \frac{N_u}{2000A_g}\right)\lambda\sqrt{f_c'}b_w d \quad \text{(see Figure 2.15)} \tag{2.29}$$

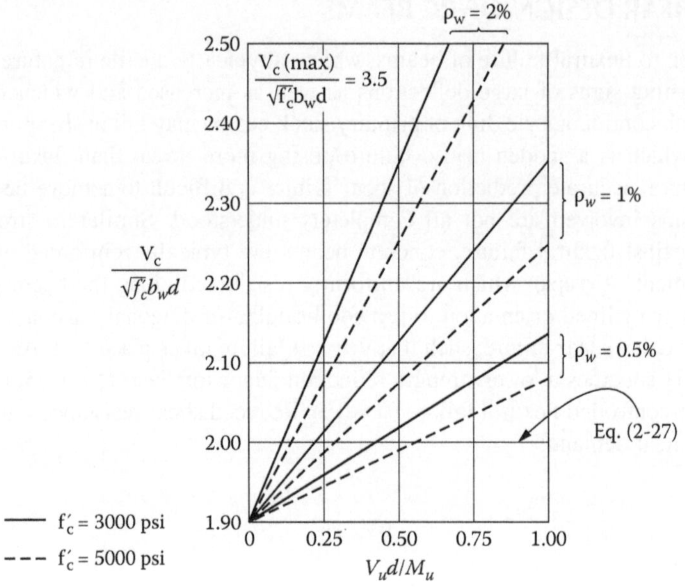

FIGURE 2.14 Concrete shear strength for shear and flexure only. (Courtesy of Portland Cement Association [2013].)

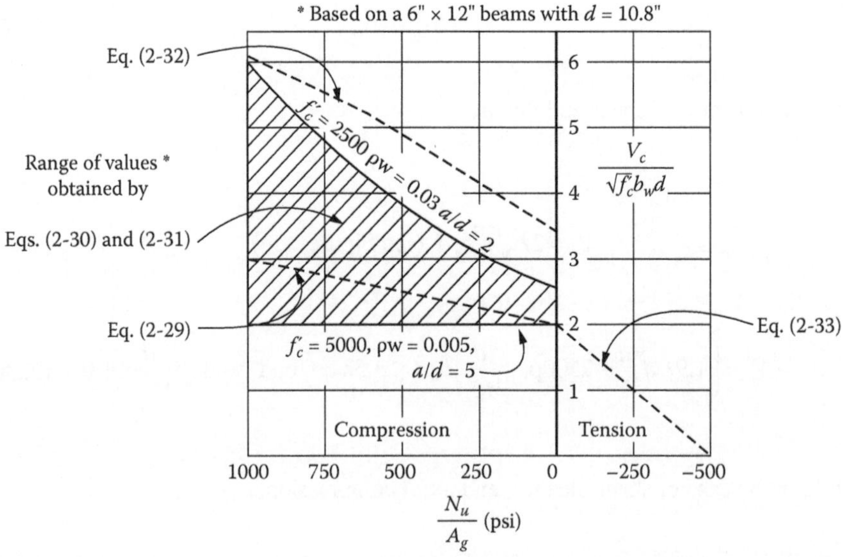

FIGURE 2.15 Concrete shear strength for shear, flexure, and axial force. (Courtesy of Portland Cement Association [2013].)

$$V_c = \left(1.9\lambda\sqrt{f_c'} + 2500\rho_w \frac{V_u d}{M_m}\right) b_w d \tag{2.30}$$

$$M_m = M_u - N_u\left(\frac{4h-d}{8}\right) \tag{2.31}$$

$$V_c \le 3.5\lambda\sqrt{f_c'} b_w d \sqrt{1 + \frac{N_u}{500 A_g}} \tag{2.32}$$

and in the case of shear, flexure, and axial tension,

$$V_c = 2\left(1 + \frac{N_u}{500 A_g}\right)\lambda\sqrt{f_c'} b_w d \ge 0 \tag{2.33}$$

where N_u is negative, N_u/A_g is in psi, and $\sqrt{f_c'} \le 100$ psi except as in Section 11.1.2.1 of ACI 318-11.

For shear strength of stirrups perpendicular to the axis of member,

$$V_s = \frac{A_v f_{yt} d}{S} \tag{2.34}$$

For shear strength of stirrups inclined with respect to the axis of member,

$$V_s = \frac{A_v f_{yt} (\sin\alpha + \cos\alpha) d}{S} \tag{2.35}$$

where S is the stirrup spacing = minimum $(\frac{d}{2}, 24")$

$$\text{If} \quad V_s > 4\sqrt{f_c'} b_w d, \qquad S = \text{minimum}\left(\tfrac{d}{4},\ 12"\right)$$

$$\text{If} \quad V_u > \phi \tfrac{V_c}{2}, \qquad \text{use } A_{v,\min}$$

with exceptions in Section 11.4.6.1 of ACI 318-11

$$A_{v,\min} = \max\left(0.75\sqrt{f_c'}\frac{b_w S}{f_{yt}}, \frac{50 b_w S}{f_{yt}}\right) \tag{2.36}$$

TABLE 2.4
Layout of Shear Design Provision of ACI 318-11.

		$V_u \le \phi V_c/2$	$\phi V_c/2 < V_u \le \phi V_c$	$\phi V_c < V_u$
Required area of stirrups, A_v		none	The larger of $0.75\sqrt{f_c'}\frac{b_w s}{f_{yt}}$ and $\frac{50 b_w s}{f_{yt}}$	The largest of $\frac{(V_u-\phi V_c)s}{\phi f_{ytd}}$, $0.75\sqrt{f_c'}\frac{b_w s}{f_{yt}}$ and $\frac{50 b_w s}{f_{yt}}$
Stirrup spacing, s	Required	–	The smaller of $\frac{A_v f_{yt}}{0.75\sqrt{f_c'} b_w}$ and $\frac{A_v f_{yt}}{50 b_w}$	The smallest of $\frac{\phi A_v f_{ytd}}{V_u-\phi V_c}$, $\frac{A_v f_{yt}}{0.75\sqrt{f_c'} b_w}$ and $\frac{A_v f_{yt}}{50 b_w}$
	Maximum	–	The smaller of $d/2$ and 24 in.	For $(V_u-\phi V_c) \le \phi 4\sqrt{f_c'} b_w d$, s is the smaller of $\frac{d}{2}$ and 24 in.
				For $\phi 4\sqrt{f_c'} b_w d < (V_u-\phi V_c) \le \phi 8\sqrt{f_c'} b_w d$, s is the smaller of $\frac{d}{4}$ and 12 in.

Source: Courtesy of Portland Cement Association (2013).

According to ACI 318-11, Section 11.4.7.9, and to avoid premature crushing of concrete struts, $V_s \le 8\sqrt{f_c'} b_w d$. The ACI 318-11 shear design procedure is summarized in Table 2.4.

Example 2.3: Design

For the simply supported beam shown in Figure 2.16, you should design for the stirrup demand along the beam, considering the ultimate load shown in the figure and knowing that

$$f_c' = 4\,ksi, f_{yt} = 50\,ksi$$

Solution:

Assume primary steel is #6
Assume stirrups are #3

$$d = 22'' - 1.5'' - \frac{3}{8} - \frac{1}{2} \times \frac{6}{8} = 19.75''$$

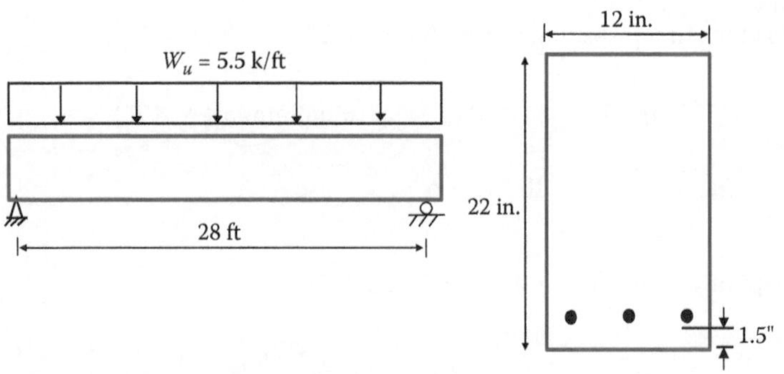

FIGURE 2.16 Beam profile and cross section for Example 2.3.

$$V_u = w_u \frac{l_n}{2} = 5.5 \times \frac{28}{2} = 77 \text{ kips at support}$$

$$V_{ud} = 77 - 5.5 \times \frac{19.75}{12} = 67.95 \text{ kips at critical section}$$

$$\phi V_c = \phi \times 2\sqrt{f_c'} b_w d = 0.75 \times 2\sqrt{4{,}000} \times 12 \times 19.75 = 22{,}483.8 \text{ lb} = 22.48 \text{ kips}$$

$$\phi V_c = 22.48\ kips < V_{ud} = 67.95 \text{ kips}$$

Therefore, shear reinforcement is required.

$$\phi V_s = V_u - \phi V_c = 67.95 - 22.48 = 45.47 \text{ kips} < \phi \times 8\sqrt{f_c'} b_w d = 89.94 \text{ kips}$$

$$S_{req'd} = \frac{\phi A_v f_{yt} d}{V_u - \phi V_c} = \frac{0.75 \times 0.22 \times 50 \times 19.75}{45.47} = 3.58 \text{ in.}$$

Try #4 stirrups

$$S_{req'd} = \frac{0.75 \times 0.4 \times 50 \times 19.75}{45.47} = 6.52 \text{ in.} \Rightarrow \text{Use 6" spacing}$$

Check $S_{\max}$:

$$\phi 4\sqrt{f_c'} b_w d = \frac{89.94}{2} = 44.97 \text{ kips} < 45.47 \text{ kips}$$

$$S_{\max} = \frac{d}{4} = 4.94 \text{ in.} < 12\text{" (the former controls)}$$

Use #4 stirrups at 4.5" o.c.only between support and a distance of 20.8"

Minimum shear reinforcement:

$$S = \frac{A_v f_{yt}}{0.75\sqrt{f_c'} b_w} = \frac{0.4 \times 50{,}000}{0.75\sqrt{4{,}000} \times 12} = 35\text{"} \le \frac{A_v f_{yt}}{50 b_w} = \frac{0.4 \times 50{,}000}{50 \times 12}$$

$$= 33\text{"} > \frac{d}{2} = 9.9 \text{ in. (controls)}$$

Determine x_c distance from support to $(V_u = \phi V_c)$

$$\phi V_c = V_u - w_u x_c \Rightarrow x_c = \frac{V_u - \phi V_c}{w_u} = \frac{77 - 22.48}{5.5} = 9.9 \text{ ft}$$

Determine x_m distance from support to $\left(V_u = \phi \frac{V_c}{2}\right)$

$$\phi \frac{V_c}{2} = V_u - w_u x_m \Rightarrow x_m = \frac{V_u - \phi \frac{V_c}{2}}{w_u} = \frac{77 - \frac{22.48}{2}}{5.5} = 11.96 \text{ ft}$$

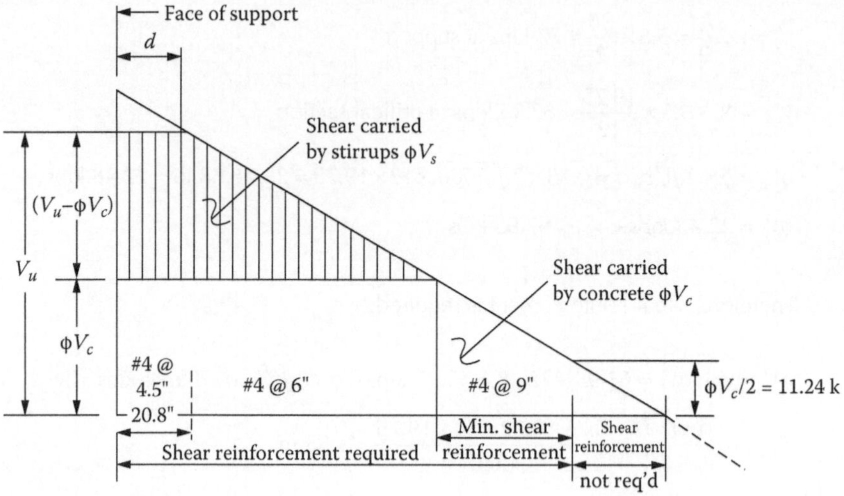

FIGURE 2.17 Stirrup design profile for Example 2.3. (Courtesy of Portland Cement Association [2013].)

The beam stirrup design is graphically presented in Figure 2.17.

Alternatively, use Table 2.5 to determine the size and spacing of stirrups

$$V_{ud} - \phi V_c = 45.47 \text{ kips}, \qquad \text{by interpolation}$$

$$\text{No. 4 stirrups @ } \frac{d}{3} = 6.6\text{" provides } \frac{36+54}{2} = 45 \text{ kips}$$

Use No. 4 stirrups at 6" o.c.

TABLE 2.5
Stirrup Shear Strength for Given Bar Size and Spacing

	Shear Strength ϕ, V_S (kips)[a]					
	No. 3 U-Stirrups		No. 4 U-Stirrups		No. 5 U-Stirrups	
Spacing	Grade 40	Grade 60	Grade 40	Grade 60	Grade 40	Grade 60
$d/2$	13.2	19.8	24.0	36.0	37.2	55.8
$d/3$	19.8	29.7	36.0	54.0	55.8	83.7
$d/4$	26.4	39.6	48.0	72.0	74.4	111.6

Source: Courtesy of Portland Cement Association (2013).

[a] Stirrups with two legs (double value for four legs, etc.).

Example 2.4: Analysis

A 9-ft-long column is subjected to the two load cases given here (Cases A and B). Check the shear reinforcement under both cases.

$$f_c' = 4\,ksi\,, f_{yt} = 50\,ksi$$

Case A:
$M_u = 67.5$ k-ft
$V_u = 15$ kips
$P_u = 150$ kips

Case B:
$M_u = 67.5$ k-ft
$V_u = 15$ kips
$P_u = 30$ kips

Solution:

Case A: $P_u = N_u = 150$ kips

$$d = 14'' - 1.5'' - 0.375 - \frac{0.875}{2} = 11.69''$$

$$\phi V_c = \phi \times 2\left(1 + \frac{N_u}{2{,}000 A_g}\right)\lambda\sqrt{f_c'}\,b_w d = 0.75 \times 2 \times \left(1 + \frac{150{,}000}{2{,}000(14 \times 12)}\right)\sqrt{4{,}000}$$
$$\times 12 \times 11.69''$$
$$= 19{,}249.26 \text{ lb} = 19.25 \text{ kips}$$

$$\phi V_c > V_u = 15 \text{ kips}$$

But $V_u > \phi \frac{V_c}{2} = 9.625$ kips, use minimum reinforcement.

Use #3 stirrups

$$S = \min\left\{\begin{array}{l} \dfrac{A_v f_{yt}}{0.75\sqrt{f_c'}\,b_w} = \dfrac{0.22 \times 50{,}000}{0.75\sqrt{4{,}000} \times 12} = 19.33 \text{ in.} \\ \dfrac{A_v f_{yt}}{50 b_w} = \dfrac{0.22 \times 50{,}000}{50 \times 12} = 18.33 \text{ in.} \\ \dfrac{d}{2} = \dfrac{11.69}{2} = 5.85 \text{ in. (controls)} \\ 24 \text{ in.} \end{array}\right\}$$

Using $S = 5.5''$ is adequate.

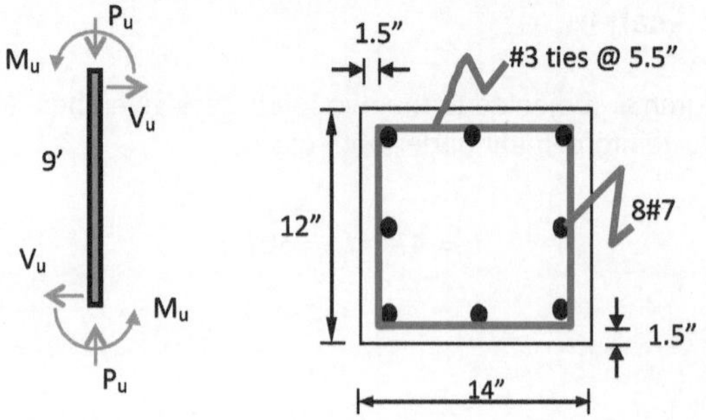

FIGURE 2.18 Column profile and cross section of Example 2.4.

Case B: $P_u = N_u = 30$ kips

$$\phi V_c = \phi \times 2\left(1+\frac{N_u}{2{,}000A_g}\right)\lambda\sqrt{f_c'}b_w d = 0.75\times 2\times\left(1+\frac{30{,}000}{2{,}000(14\times 12)}\right)\sqrt{4{,}000}$$

$$\times 12\times 11.69''$$

$$= 14{,}496.36 \text{ lb} = 14.5 \text{ kips}$$

$$\phi V_c < 15 \text{ kips}$$

$$S_{max} = \frac{d}{2} = 5.85 \text{ in. (controls)}$$

$$\phi V_n = \phi V_c + \phi V_s = 14.5 + \frac{0.75\times 0.22\times 50\times 11.69}{5.5} = 32.04 \text{ kips} > V_u = 15 \text{ kips}$$

2.4 INTERNAL REINFORCEMENT TO CONFINE RC COLUMNS

ACI 318-11 does not rely on increasing strength by means of core confinement using internal reinforcement. However, as the load and deformation increase to an extent that spalls off the concrete cover under axial compression, the code tries to compensate for loss of strength from the spalled cover by introducing lateral reinforcement that increases the strength by a comparable amount for concentrically loaded columns. Since this confinement effect may only be effectively achieved in circular cross sections, ACI 318-11 limits that provision to spirally reinforced columns (Section 10.9.3).

$$\rho_s = 0.45\left(\frac{A_g}{A_{ch}} - 1\right)\frac{f_c'}{f_{yt}}$$

Rectangular and square columns are designed for ties using imposed shear forces, but no special confinement provision is added in this case. As a minimum, ACI 318-11 requires tie spacing to be the minimum of 16 times the bar diameter, 48 times the tie diameter, or the least dimension of the column (Section 7.10.5.2)

Example 2.5: Design

Design for minimum tie spacing in the following column:

$f_c' = 4{,}000$ psi

$f_y = 60{,}000$ psi

$f_{yt} = 50{,}000$ psi

Solution:

$$\rho = \frac{4 \times 0.6}{15 \times 15} = 0.0107 > 0.01 \qquad \text{O.K.}$$

Use #3 ties

$$\text{Spacing} = \min\left\{\begin{array}{ll} 16d_b = 16 \times \frac{7}{8} = 14" & \text{controls} \\ 48d_t = 48 \times \frac{3}{8} = 18", & \\ 15" & \end{array}\right\}$$

Thus, use #3 ties at 14" c/c.

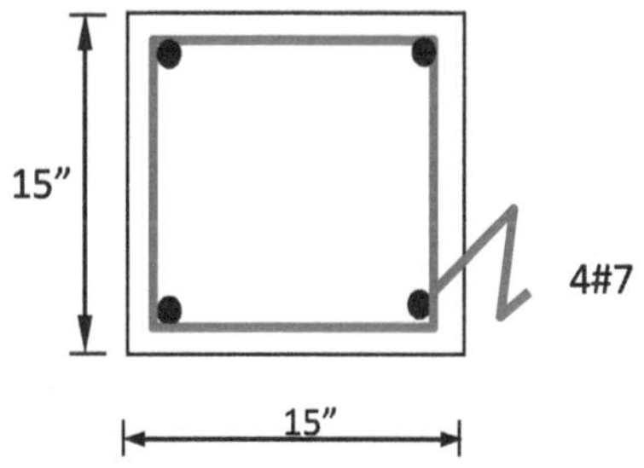

FIGURE 2.19 Square column cross section of Example 2.5.

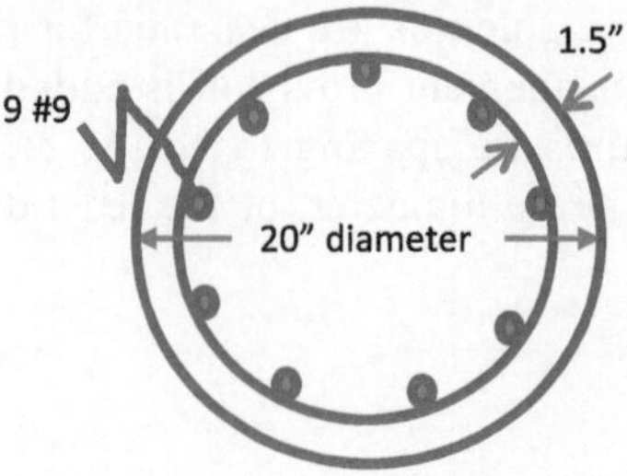

FIGURE 2.20 Circular column cross section of Example 2.6.

Example 2.6: Design

Design for spiral reinforcement in the following circular column using ACI 318-11:

$f'_c = 4{,}000$ psi

$f_y = 60{,}000$ psi

$f_{yt} = 50{,}000$ psi

Solution:

$$\rho = \frac{9 \times 1}{\frac{\pi}{4} \times 20^2} = 0.0286 > 0.01 \qquad \text{O.K.}$$

Use #3 spirals

$$\rho_s = 0.45\left(\frac{A_g}{A_{ch}} - 1\right)\frac{f'_c}{f_{yt}} = 0.45\left(\frac{\frac{\pi}{4} \times 20^2}{\frac{\pi}{4} \times (20-3)^2} - 1\right) \times \frac{4}{50} = 0.0138$$

$$\rho_s = \frac{4A_{sp}}{d_c S} \Rightarrow S = \frac{4 \times 0.11}{(20-3) \times 0.0138} = 1.87 \text{ in.} > 1 \text{ in.}$$

$$< 3 \text{ in.}$$

Use #3 spiral @ $1\frac{3}{4}$ in.

Confinement models for increasing core strength (justifying ACI 318-11 equation):

Circular Columns

$$f_l = \frac{1}{2}\rho_s f_{yt}$$

Due to the arch action in between hoops and spirals

$$f_l' = \frac{1}{2}k_e\rho_s f_{yt}$$

$$\rho_s = \frac{4A_{sp}}{d_c S}$$

For circular hoops

$$k_e = \frac{\left(1-\frac{s'}{2d_c}\right)^2}{1-\rho_{cc}}$$

For circular spirals

$$k_e = \frac{1-\frac{s'}{2d_c}}{1-\rho_{cc}}$$

where
s' = clear spacing of hoops or clear pitch of spiral
d_c = diameter of concrete core c/c
ρ_{cc} = ratio of longitudinal reinforcement area to the area of core

$$f_{cc}' = f_c'\left(-1.254 + 2.254\sqrt{1+\frac{7.94f_l'}{f_c'}} - 2\frac{f_l'}{f_c'}\right) \qquad \text{Mander } et\ al. (1988)$$

ACI 318-11 equation of Section 10.9.3:

$$\rho_s = 0.45\left(\frac{A_g - A_c}{A_c}\right)\frac{f_c'}{f_{yt}}$$

$$\rho_s f_{yt} A_c = 0.45\,f_c' A_{cover}$$

$$2\,f_l A_c = 0.45\,f_c' A_{cover}$$

$$f_{cc}' = f_c' + 4.1 f_l \qquad \text{Richart } et\ al. (1928)$$

$$f_l = \frac{f_{cc}' - f_c'}{4.1}$$

$$0.488(f_{cc}' - f_c')A_c = 0.45 f_c' A_{cover}$$

$$0.57 \times 0.85(f_{cc}' - f_c')A_c = 0.53 \times 0.85 f_c' A_{cover}$$

$$0.85(f_{cc}' - f_c')A_c = 0.92 \times 0.85 f_c' A_{cover}$$

The two terms on both sides of this equation are almost equal.

2.5 SERVICE LOAD CALCULATIONS IN BEAMS

Even though the working-design method is something of the past, service-load calculations are needed when designing for serviceability conditions like deflections and cracking. This serviceability calculation requires checking the cracking moment and, in most cases, carrying out computations under actual service loads, assuming linear elastic theory and cracked sections. It is worth emphasizing that linear elastic computations are valid until the extreme fiber compressive concrete stress exceeds $0.7f'_c$ (Park and Paulay 1975; Charkas, Rasheed, and Melhem 2003). Accordingly, cracked-section analysis under service loads is primarily performed with linear elastic analysis except for very heavily reinforced sections (Charkas, Rasheed, and Melhem 2003).

Example 2.7: Calculate

For the beam loaded by four-point bending shown here (Figure 2.21):

1. Calculate the cracking moment (M_{cr})
2. Calculate the steel and concrete stresses for $P = 30$ kN
3. Calculate the steel and concrete stresses for $P = 60$ kN

Assume $f'_c = 25$ MPa and $f_y = 414$ MPa

Solution:

Assume ϕ10 stirrups

$$d = 700 - 38 - 10 - (22/2) = 641 \text{ mm}$$

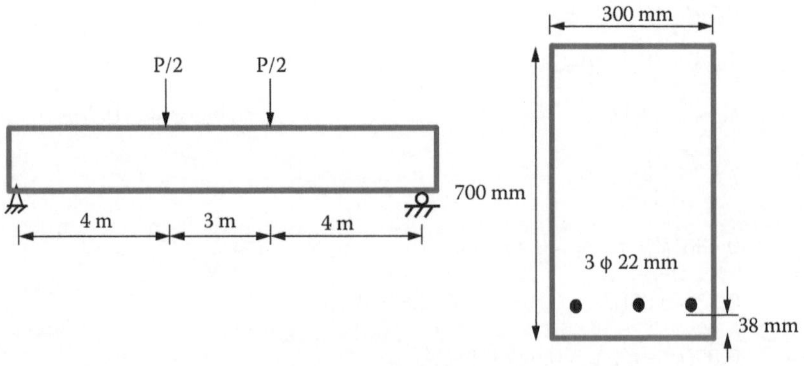

FIGURE 2.21 Beam profile and cross section for Example 2.7.

1. Cracking moment:

$$n = \frac{E_s}{E_c} = \frac{200{,}000}{4{,}700\sqrt{25}} = 8.51$$

$$A_s = \frac{\pi}{4} \times 22.225^2 \times 3 = 1164 \text{ mm}^2 \text{(Table A-1)}$$

$$\bar{y} = \frac{bh\frac{h}{2} + (n-1)A_s d}{bh + (n-1)A_s} = \frac{300 \times 700 \times 350 + 7.5 \times 1164 \times 641}{300 \times 700 + 7.5 \times 1164} = 362 \text{ mm}$$

$$I_{gt} = \frac{bh^3}{12} + bh\left(\frac{h}{2} - \bar{y}\right)^2 + (n-1)A_s(d-\bar{y})^2$$

$$= \frac{300 \times 700^3}{12} + 300 \times 700 \times (350 - 362)^2 + 7.5 \times 1164 \times (641 - 362)^2$$

$$= 9.285 \times 10^9 \text{ mm}^4 = 0.009285 \text{ m}^4$$

$$f_r = 0.62\sqrt{f_c'} = 0.62\sqrt{25} = 3.1 \text{ MPa}$$

$$f_r = \frac{M_{cr}(h - \bar{y})}{I_{gt}} \Rightarrow M_{cr} = \frac{3.1 \times 0.009285}{(700 - 362) \times 10^{-3}} = 0.0852 \text{ MN-m} = 85.2 \text{ kN-m}$$

2. Stresses at $P = 30$ kN

$$M_{max} = 15 \times 4 = 60 \text{ kN-m} < 85.2 \text{ kN-m} \Rightarrow \text{uncracked section}$$

$$f_{cc} = \frac{M\bar{y}}{I_{gt}} = \frac{60 \times 362 \times 10^{-3}}{0.009285} = 2339.26 \text{ kPa} = 2.34 \text{ MPa} \approx 9.4\% f_c'$$

$$f_{ct} = \frac{M(h - \bar{y})}{I_{gt}} = \frac{60 \times (700 - 362) \times 10^{-3}}{0.009285} = 2184.17 \text{ kPa} = 2.18 \text{ MPa} \approx 70\% f_r$$

$$f_s = n\frac{M(d - \bar{y})}{I_{gt}} = 8.5 \times \frac{60 \times (641 - 362) \times 10^{-3}}{0.009285} = 15{,}324.72 \text{ kPa}$$

$$= 15.32 \text{ MPa} \approx 3.7\% f_y$$

3. Stresses at $P = 60$ kN

$$M_{max} = 30 \times 4 = 120 \text{ kN-m} > 85.2 \text{ kN-m}$$

Cracked section

Assume elastic behavior

$$bkd\frac{kd}{2} = nA_s(d - kd)$$

$$\frac{1}{2}k^2 = n\frac{A_s}{bd}(1 - k) \Rightarrow k = \sqrt{2n\rho + (\overline{n\rho})^2} - n\rho$$

$$\rho = \frac{1164}{300 \times 641} = 0.006053$$

$$k = 0.273 \Rightarrow j = 1 - \frac{k}{3} = 0.909$$

$$kd = 0.273 \times 641 = 175 \text{ mm}$$

$$M = \frac{1}{2} f_{cc} kdbjd$$

$$f_{cc} = \frac{M}{\frac{1}{2} bd^2 kj} = \frac{120}{\frac{1}{2} \times 300 \times 641^2 \times 10^{-9} \times 0.273 \times 0.909}$$

$$= 7846 \text{ kPa} = 7.85 \text{ MPa} \approx 31\% f_c' < 70\% f_c' \qquad \text{linear O.K.}$$

$$M = A_s f_s jd$$

$$f_s = \frac{M}{A_s jd} = \frac{120}{1164 \times 10^{-6} \times 0.909 \times 641 \times 10^{-3}}$$

$$= 176{,}932 \text{ kPa} = 176.93 \text{ MPa} \approx 43\% f_y$$

Chapter Problems

Problem 2.1

Use the stress–strain curve shown below to determine α and γ for different values of f_c' (3–8 ksi). Assume $f_c'' = f_c'$. Determine the results using β1 from ACI 318 equation and Park and Paulay's model. Please realize that the latter is only approximate, since it is derived for the entire parabola. Hint: Integrate the area under the curve.

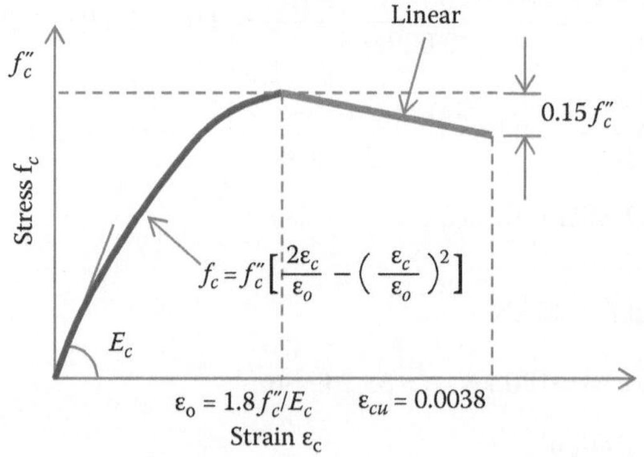

FIGURE 2.P.1

Problem 2.2

Repeat Problem 2.1 using the original Hognestad's parabola, that was derived in this chapter. However, use Park and Paulay's derivation of the actual β_1 instead of the ACI 318 equation.

Problem 2.3

For the beam designed in Example 2.2, determine the actual moment capacity of the doubly reinforced section (M_u).

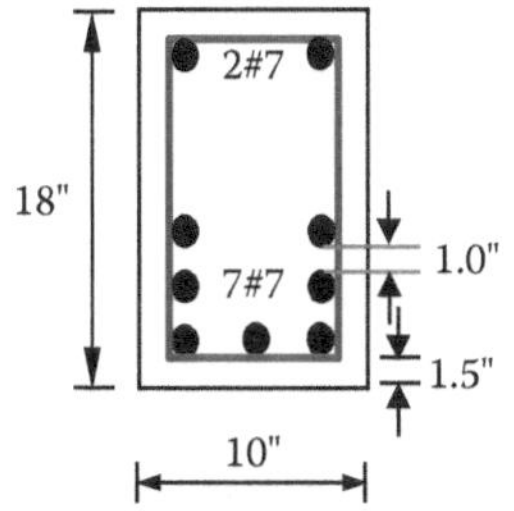

FIGURE 2.P.2

Problem 2.4

Design the following rectangular section as a doubly reinforced beam assuming the strain limit of 0.005 for tension-controlled sections. M_u = 60 k-ft and clear cover is 1.0 in. Assume #3 bars for stirrups and compression steel.

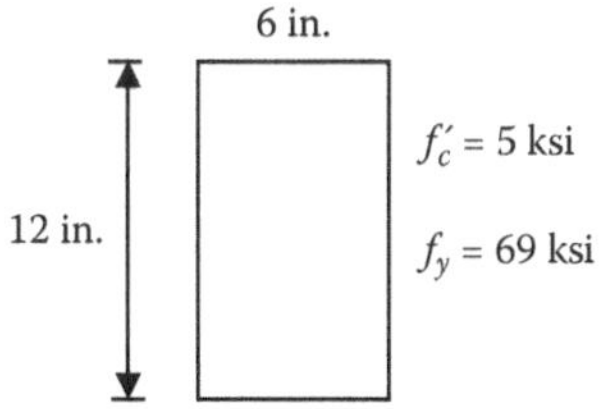

FIGURE 2.P.3

Problem 2.5

For the beam section in Problem 2.3, design the beam to resist shear such that it fails in flexure and not shear. The beam is under its own weight in addition to a single concentrated live load at mid span, as shown.

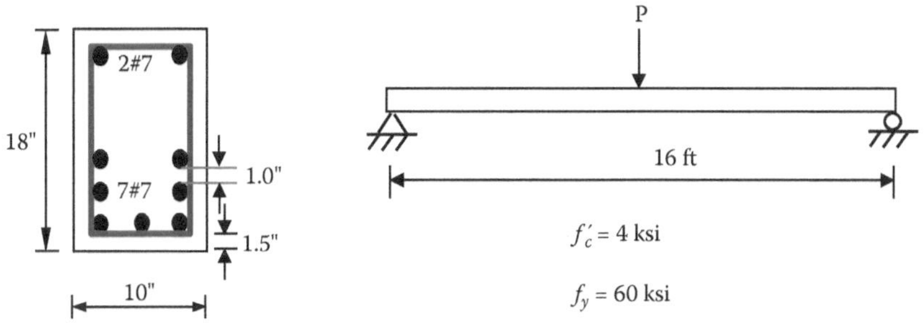

FIGURE 2.P.4

Problem 2.6

For the beam given in Example 15.3 of the ACI440.2R-08, design the beam in shear for the original dead and live loads. Then, check if your design provides adequate stirrup reinforcement under the upgraded loads. Otherwise, indicate the need for shear strengthening.

Problem 2.7

For the beam given in Problem 2.6, check extreme fiber stresses and steel stress under (a) dead load only and (b) dead load + 20% live load at the time of strengthening. Determine ε_{bi} at mid span in both cases.

Problem 2.8

For the circular column section in Example 2.6, determine the increase in column core capacity relative to the loss of concrete cover using the Mander et al. (1988) model. Take the #3 spiral transverse reinforcement at

(a) $S = 1.75$ in. as obtained based on the ACI 318 equation
(b) $S = 1$ in. to invoke higher confinement effect

Problem 2.9

Repeat the third part of Example 2.7 by deriving I_{cr} and determining the stresses based upon it. Compare the answers to what you got in Example 2.7. What do you conclude?

APPENDIX A

TABLE A-1

Rebar is Sized Nominally by "Eigths of an Inch" of the Bar's Diameter. A #3 Bar is 3/8" in Diameter. A #6 Bar is 3/4" in Diameter. 3/4 is the Same as 6/8. Every Bar Can be Measured in Eights of an Inch. A #10 Bar is About 10/8″ in Diameter. This is the Same as 1.27"

	Diameter	Area(in²)	Lbs/Inft	Diameter (Metric)	Metric Bar Size
No. 2	0.250"	0.05	0.167 lbs	6.35 mm	No. 6
No. 3	0.375"	0.11	0.376 lbs	9.52 mm	No. 10
No. 4	0.500"	0.20	0.668 lbs	12.7 mm	No. 13
No. 5	0.625"	0.31	1.043 lbs	15.8 mm	No. 16
No. 6	0.750"	0.44	1.502 lbs	19.05 mm	No. 19
No. 7	0.875"	0.60	2.044 lbs	22.225 mm	No. 22
No. 8	1.000"	0.79	2.670 lbs	25.4 mm	No. 25
No. 9	1.128"	1.00	3.400 lbs	28.65 mm	No. 29
No. 10	1.270"	1.27	4.303 lbs	32.25 mm	No. 32
No. 11	1.410"	1.56	5.313 lbs	35.81 mm	No. 36
No. 14	1.693"	2.25	7.650 lbs	43.0 mm	No. 43
No. 18	2.257"	4.00	13.60 lbs	57.33 mm	No. 57

TABLE A-2
A Few Years Ago, the Metric Equivalents Were Given Nominal Size Names, But They Were Redesignated. It Used to Run 5, 10, 15, 20 on to 55 in Increments of 5. But Then Someone Decided Otherwise

Here's what they used to be called:

Imperial Size	Former Metric Name	Current Metric Name
No. 2	No. 5	No. 6
No. 3	No. 10	No. 10
No. 4	not used	No. 13
No. 5	No. 15	No. 16
No. 6	No. 20	No. 19
No. 7	not used	No. 22
No. 8	No. 25	No. 25
No. 9	No. 30	No. 29
No. 10	not used	No. 32
No. 11	No. 35	No. 36
No. 14	No. 45	No. 43
No. 18	No. 55	No. 57

REFERENCES

ACI 318-11. 2011. Building code requirements for structural concrete and commentary. ACI Committee 318. Farmington Hills, MI.

ACI SP-55. 1978. Detroit: American Concrete Institute.

Beer, F. P., and E. R. Johnston Jr. 1992. *Mechanics of materials*. 2nd ed. New York: McGraw-Hill.

Charkas, H., H. A. Rasheed, and H. G. Melhem. 2003. Rigorous procedure for calculating deflections of FRP-strengthened RC beams. *ACI Structural Journal* 100 (4): 529–39.

Hognestad, E. 1951. *A study of combined bending and axial load in reinforced concrete members*. University of Illinois Engineering Experiment Station Bulletin Series No. 399. University of Illinois, Urbana.

Kaar, P. H., N. W. Hanson, and H. T. Capell. 1978. Stress-strain characteristics of high strength concrete. In *ACI Publication SP-55,* 161–86. Detroit: American Concrete Institute.

MacGregor, J. G. 1992. *Reinforced concrete: Mechanics & design*. 2nd ed. Upper Saddle River, NJ: Prentice Hall.

Mander, J. B., M. J. N. Priestley, and R. Park. 1988. Theoretical stress-strain model for confined concrete. *Journal of Structural Engineering, ASCE*, 114 (8), 1827–49.

Nilson, A. H. 1997. *Design of concrete structures*. 12th ed. New York: WCB/McGraw-Hill.

Park, R., and T. Paulay. 1975. *Reinforced concrete structures*. New York: John Wiley and Sons.

Portland Cement Association. 2013. *Notes on ACI 318-11 building code requirements for structural concrete with design applications*. 12th ed. Ed. M. E. Kamara and L. C. Novak. Skokie, IL Portland Cement Association.

Richart, F. E., A. Brandtzaeg, and R. L. Brown. 1928. A study of the failure of concrete under combined compressive stresses. Bulletin 185, University of Illinois Engineering Experiment Station, University of Illinois, Urbana.

Rüsch, H. 1960. Research toward a general flexural theory for structural concrete. *ACI Journal, Proceedings* 57 (1): 1–28.

3 Constituent Materials and Properties

3.1 OVERVIEW

Fiber-reinforced polymer (FRP) is composed of two material phases: fiber and polymer matrix. Fibers are impregnated into the polymer matrix to form a macroscopically orthotropic layer of material with distinctly higher mechanical properties along the fiber direction compared to the transverse directions. The advantages of using FRP are the high strength and stiffness-to-weight ratio along the fiber direction, ease of application in construction due to its light weight, corrosion resistance, electromagnetic inertness, and design versatility in which high strength and stiffness (fibers) may be oriented where needed in design. Continuous fibers become extremely strong and stiff as fiber diameter becomes smaller due to the reduction and sometimes elimination of defects in the microstructure, as illustrated in Figure 3.1. On the other hand, small-diameter fibers are not capable of carrying axial compression or shear stresses due to the lack of shear transfer medium between them. Thus, the fibers are embedded into a polymeric matrix that binds them together and allows load transfer by shear among the fibers. Additional specifics about fibers and matrix are described in the following sections.

3.2 FIBERS

Fibers are typically made of glass, carbon, and aramid. Other synthetic fibers are made of polymers which are not used in structural applications due to their low mechanical properties. Glass fibers are primarily composed of silicon dioxide with some modifying agents (Gibson 1994). E-glass (electrical glass) accounts for the largest production of glass fibers in industry due to its low cost despite its mechanical properties that are lower than other grades of glass fibers, as shown in Figure 3.2. On the other hand, S-glass (structural glass) is more expensive to produce, but it has significantly higher strength and slightly higher modulus, as shown in Table 3.1 (Gibson 1994) and Table 3.2 (Hyer 1998). C-glass (chemical glass) has an improved durability against alkali and acid attacks.

Carbon fibers or graphite fibers are the most widely used fibers in industry due to their high stiffness and strength as well as environmental stability. Carbon fibers contain less than 95% carbon, while graphite fibers have at least 99% carbon (Schwartz 1984). Graphite and carbon fibers are still the most expensive fibers on the market,

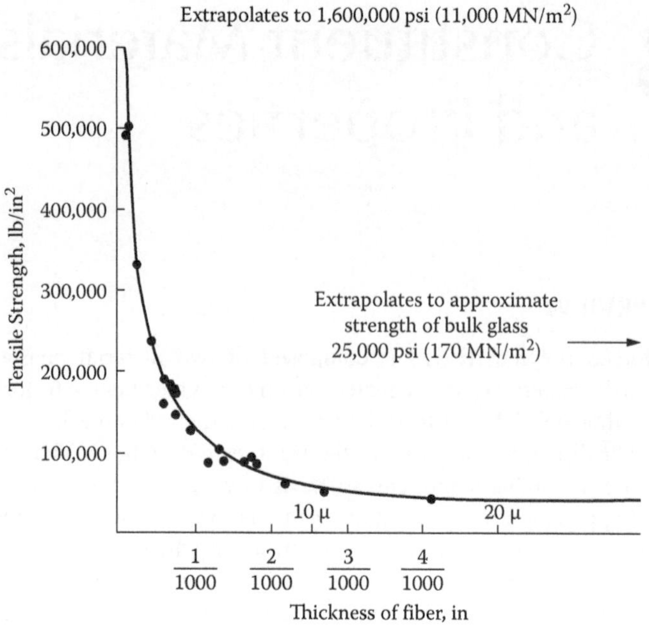

FIGURE 3.1 Glass fiber strength in tension vs. glass fiber thickness. (Courtesy of Griffith [1920] and Gordon [1976].)

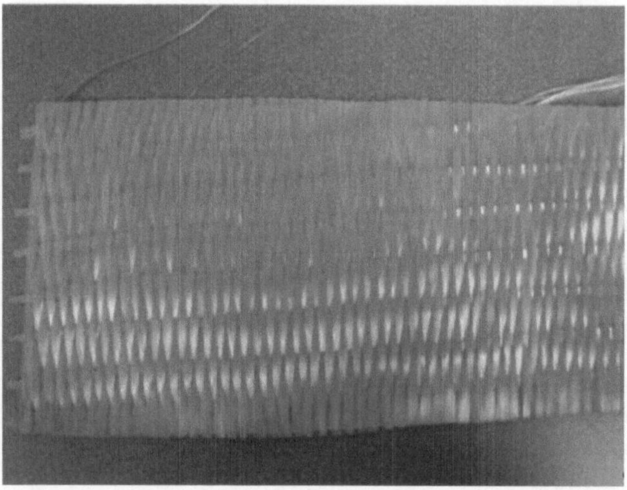

FIGURE 3.2 Unidirectional E-glass fiber sheet used in concrete strengthening.

TABLE 3.1
Comparison of Fiber Properties and Selected Metal

Material	Tensile Strength, 10^3 psi (MPa)	Tensile Modulus, 10^6 psi (GPa)	Density, lb/in.^{3} (g/cm^3)
Bulk 6061T6 Aluminum	45.0 (310)	10.0 (69)	0.098 (2.71)
Bulk SAE 4340 Steel	150.0 (1034)	29.0 (200)	0.283 (7.83)
E-glass fibers	500.0 (3448)	10.5 (72)	0.092 (2.54)
S-glass fibers	650.0 (4482)	12.5 (86)	0.090 (2.49)
Carbon fibers (PAN precursor)			
AS-4 (Hercules)	580.0 (4000)	33.0 (228)	0.065 (1.80)
IM-7 (Hercules)	785.0 (5413)	40.0 (276)	0.064 (1.77)
T-300 (Amoco)	530.0 (3654)	33.5 (231)	0.064 (1.77)
T-650/42 (Amoco)	730.0 (5033)	42.0 (290)	0.064 (1.77)
Carbon fibers (pitch precursor)			
P-55 (Amoco)	250.0 (1724)	55.0 (379)	0.072 (1.99)
P-75 (Amoco)	300.0 (2068)	75.0 (517)	0.072 (1.99)
P-100 (Amoco)	325.0 (2241)	100.0 (690)	0.078 (2.16)
Aramid fibers			
Kevlar® 29 (Dupont)	550.0 (3792)	9.0 (62)	0.052 (1.44)
Kevlar® 49 (Dupont)	550.0 (3792)	19.0 (131)	0.053 (1.47)
Boron fibers			
0.004" diameter (Textron)	510.0 (3516)	58.0 (400)	0.093 (2.57)
0.0056" diameter (Textron)	510.0 (3516)	58.0 (400)	0.090 (2.49)
Silicon carbide fibers			
0.0056" diameter (Textron)	500.0 (3448)	62.0 (427)	0.110 (3.04)

Source: Courtesy of Gibson (1994).

TABLE 3.2
Properties of Common Glass Fibers

	Glass Type		
Property	**E**	**C**	**S**
Diameter (μm)	8–14	—	10
Density (kg/m^3)	2540	2490	2490
Tensile modulus (GPa)	72.4	68.9	85.5
Tensile strength (MPa)	3450	3160	4590
Elongation (%)	1.8–3.2	4.8	5.7
Coeff. of thermal expansion ($\times 10^{-6}$/°C)	5.0	7.2	5.6
Thermal conductivity (W/m/°C)	1.3	—	—
Specific heat (J/kg/°K)	840	780	940

Source: Courtesy of Hyer (1998).

FIGURE 3.3 Carbon fiber yarn and carbon and glass fiber sheets used in strengthening.

despite the considerable drop in their prices, because of the high demand for these fibers (Reinhart 1990), as seen in Figure 3.3. Aramid polymeric fibers, which also have the trade name of Kevlar®, are used in structural applications as well. Aramid has approximately half the density of glass with very high strength, toughness, ductility, and impact resistance (Gibson 1994).

Boron fibers are composites made from coating a substrate of carbon or tungsten with boron. They are as heavy as glass, as demonstrated in Table 3.1 (Gibson 1994), and expensive to produce.

3.3 MATRIX

The matrix in a composite plays various roles such as holding the fibers into the composite part shape, protecting fibers from direct exposure to the environment, transferring the stresses through the fiber–matrix interface to the fibers, and resisting some of the applied load, especially transverse normal stresses and interlaminar shear stresses (Barbero 2011). The application of a composite is limited by the properties of its matrix. The thermal stability and useable service temperature as well as chemical resistance, moisture resistance, and abrasion resistance are all dependent on the matrix and its properties. Certain conditions, such as moisture, act on lowering the glass transition temperature (T_g) of the polymer matrix, thus significantly degrading the composite when the operational temperature exceeds T_g.

The matrix transitions from its operational state, where it is stiff and glassy, to a soft rubbery state once T_g is exceeded (Hyer 1998). In general, matrix materials can be made of polymers or resins, metals, or ceramics. The polymer matrix is the most common among matrix materials because of the ease of manufacturing complex components and relatively inexpensive tooling (Barbero 2011). This text focuses on polymer matrices, since they are widely used to strengthen concrete structures. Resins or polymers are classified into thermosetting and thermoplastic matrices.

3.3.1 Thermosetting Resins

Thermosetting resins are formed into a polymer matrix through an irreversible cross-linking chemical process called *resin curing*. These resins are the most commonly used polymer system. This is due to the fact that they have low viscosity, allowing thorough fiber immersion, and they require low processing temperatures and short processing time. They also cost less than thermoplastic resins (Hyer 1998). Resin shelf life refers to the amount of time that a resin system can be stored without degradation prior to mixing (Barbero 2011). Resin pot life is the time span during which mixed resin is still workable and applicable (Barbero 2011). The properties of various widely used thermosetting resins are listed in Table 3.3 (Hyer 1998).

TABLE 3.3
Properties of Thermosetting Polymers at Room Temperature

	Thermosetting Polymer				
Property	**Polyester**	**Vinyl Ester**	**Epoxy**	**Bismaleimide**	**Polyimide**
Density (kg/m^3)	1100–1500	1150	1100–1400	1320	1430–1890
Tensile modulus (GPa)	1.2–4.5	3–4	2–6	3.6	3.1–4.9
Shear modulus (GPa)	0.7–2	—	1.1–2.2	1.8	—
Tensile strength (MPa)	40–90	65–90	35–130	48–78	70–120
Compressive strength (MPa)	90–250	127	100–200	200	—
Elongation (%)	2–5	1–5	1–8.5	1–6.6	1.5–3
Coeff. of thermal expansion (×10^{-6}/°C)	60–200	53	45–70	49	90
Thermal conductivity (W/m/°C)	0.2	—	0.1–0.2	—	—
Specific heat (J/kg/K)	—	—	1250–1800	—	—
Glass transition temperature (°C)	50–110	100–150	50–250	250–300	280–320
Water absorption (%) [24h @ 20°C]	0.1–0.3	—	0.1–0.4	—	0.3
Shrinkage on curing (%)	4–12	1–6	1–5	—	—

Source: Courtesy of Hyer (1998).

3.3.2 Thermoplastic Resins

Thermoplastic resins obtain their mechanical properties through entanglement of the polymer chains; thus, they do not produce any cross-linking chemical process. The entanglement process is reversible, so upon heating, the chains disentangle, causing the polymer to turn to a viscous fluid. Upon cooling, the resin solidifies into a plastic shape. Thermoplastic resins are very viscous at processing temperatures, easily causing damage to long fibers. It is important to note that thermoplastic resins have no limits on shelf and pot life, giving them an advantage in this respect (Barbero 2011). In addition, composites made of these resins are easy to repair by heating individual parts to their soft state, reshaping them, and then cooling them off (Hyer 1998). However, thermoplastic resins are more expensive than thermosetting resins and require higher energy to form. Some of the widely used thermoplastic resins are listed in Table 3.4 (Hyer 1998).

3.4 FIBER AND COMPOSITE FORMS

For infrastructure-strengthening applications, fibers or composites are available in various forms. The most widely used forms are

1. *Uniaxial fiber sheets*: These are sheets of dry continuous fibers in their simplest unprocessed state. These sheets are the most commonly used fiber forms with hand lay-up processing, in which the resin is manually added to make the composite layers (Arduini, Tommaso, and Nanni 1997; Rasheed et al. 2010), as illustrated in Figure 3.4.
2. *2-D fabrics*: These are 2-D (two-dimensional) textile fabrics of continuous uniaxial fibers along with woven fibers in the transverse direction. Because of the weaving angle, the mechanical properties are slightly lower than those of sheets with higher variability in the properties of the manufactured fabric composites because of the existence of resin-rich regions at the weave locations. It is also common to use the hand lay-up process with the fabric (Bencardino, Spadea, and Swamy 2002), as seen in Figure 3.5.
3. *Prefabricated plates*: These are laminated plates manufactured from preimpregnated (prepreg) tapes or sheets. The prepreg lay-up process may include autoclave oven curing with heat, pressure, and suction to improve the quality of the final product (raise the T_g and reduce the voids; Tamimi et al. 2011), as seen in Figure 3.6.
4. *Pultruded FRP bars and tape*: These are pultruded FRP rounded bars or flat FRP tape with a typical 60% fiber content by volume. They are used as resin-bonded near-surface-mounted (NSM) bars or tape available as a final product (Alkhrdaji, Nanni, and Mayo 2000; Rasheed et al. 2010), as seen in Figure 3.6–3.7.

3.5 ENGINEERING CONSTANTS OF A UNIDIRECTIONAL COMPOSITE LAMINA

Unlike isotropic materials that have similar properties in all directions, a uniaxial lamina is orthotropic, with distinct properties along the fiber, transverse, and through-the-thickness directions, as seen in Figure 3.8. Isotropic materials have

TABLE 3.4
Properties of Thermoplastic Polymers at Room Temperature

	Aromatic Polyimides							
Property	**Polyetherimide (Ultem)**	**Polyamide-Imide (Torlon)**	**PPS: Polyphenylene Sulfide**	**PEEK: Polyether Ether Ketone**	**PS: Polysulfone**	**PP: Polypropylene**	**Nylon: Nylon 6, 6**	**PC: Polycarbonate**
Density (kg/m^3)	1270	1400	1340	1320	1240	900	1140	1060–1200
Tensile modulus (GPa)	3	5	3.3	—	2.5	1–1.4	1.4–2.8	2.2–2.4
Tensile strength (MPa)	105	95–185	70–75	92–100	70–75	25–38	60–75	45–70
Compressive strength (MPa)	140	276	110	—	—	—	34	86
Elongation (%)	60	12–18	3	150	50–100	300	40–80	50–100
Coeff. of thermal expansion ($\times 10^{-6}/°C$)	62	36	54–100	—	56–100	110	90	70
Thermal conductivity (W/m/°C)	—	—	—	—	—	0.2	0.2	0.2
Glass transition temperature (°C)	217	243–274	85	143	190	−20 to −5	50–60	133
Water absorption (%) [24h @ 20°C]	0.25	0.3	0.2	0.1	0.2	0.03	1.3	0.1

Source: Courtesy of Hyer (1998).

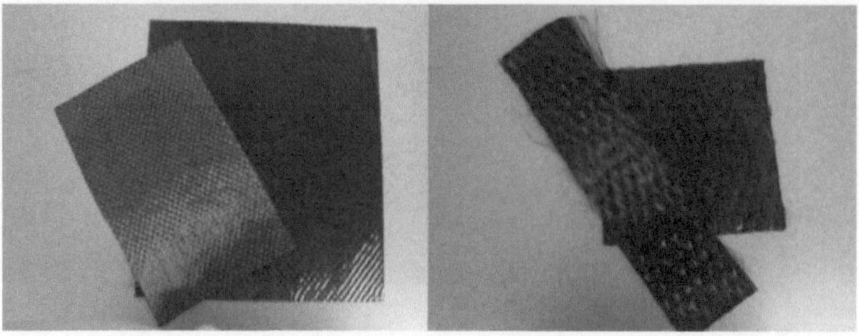

FIGURE 3.4 Carbon fiber dry sheets and composite laminas used in strengthening.

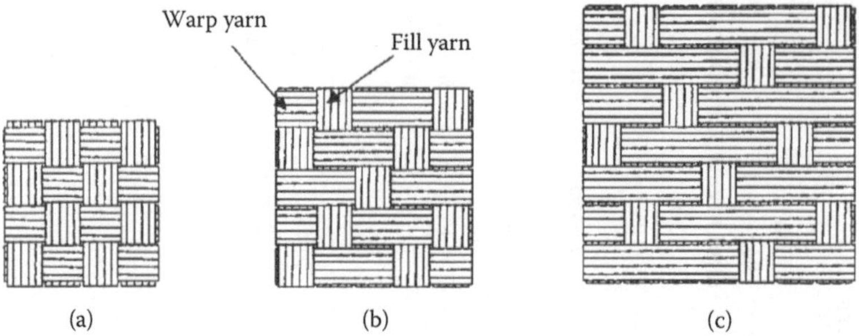

FIGURE 3.5 Examples of biaxial woven fabrics. (Courtesy of Barbero [2011].)

FIGURE 3.6 Carbon and glass prefabricated tape and plate used in strengthening.

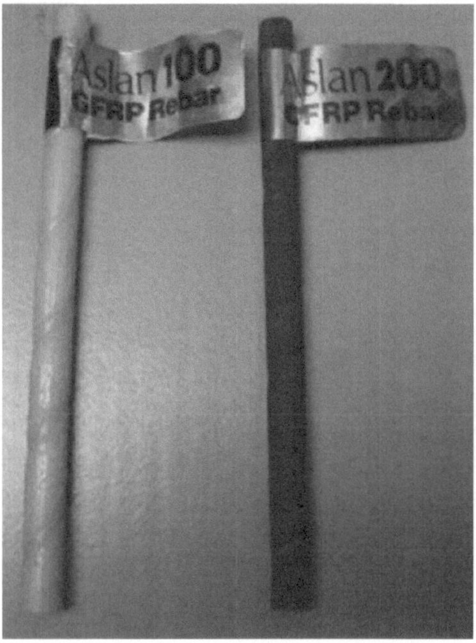

FIGURE 3.7 Carbon and glass FRP bars used in NSM strengthening technology.

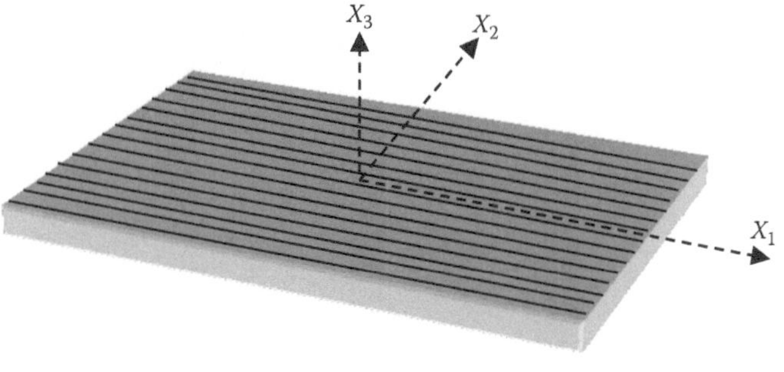

FIGURE 3.8 FRP lamina with the principal material directions.

only two independent engineering constants, which are Young's modulus of elasticity (E) and Poisson's ratio (ν). Conversely, orthotropic laminas have nine distinct engineering parameters, including three Young's moduli along the three principal materials directions (E_1, E_2, E_3), three independent Poisson's ratios (ν_{12}, ν_{13}, ν_{23}), and three shear moduli (G_{12}, G_{13}, G_{23}). The generalized 3-D compliance relationship of an orthotropic sheet is

$$\begin{Bmatrix} \varepsilon_{11} \\ \varepsilon_{22} \\ \varepsilon_{33} \\ \gamma_{12} \\ \gamma_{13} \\ \gamma_{23} \end{Bmatrix} = \begin{bmatrix} \frac{1}{E_1} & -\frac{\nu_{21}}{E_2} & -\frac{\nu_{31}}{E_3} & 0 & 0 & 0 \\ -\frac{\nu_{12}}{E_1} & \frac{1}{E_2} & -\frac{\nu_{32}}{E_3} & 0 & 0 & 0 \\ -\frac{\nu_{13}}{E_1} & -\frac{\nu_{23}}{E_2} & \frac{1}{E_3} & 0 & 0 & 0 \\ 0 & 0 & 0 & \frac{1}{G_{12}} & 0 & 0 \\ 0 & 0 & 0 & 0 & \frac{1}{G_{13}} & 0 \\ 0 & 0 & 0 & 0 & 0 & \frac{1}{G_{23}} \end{bmatrix} \begin{Bmatrix} \sigma_{11} \\ \sigma_{22} \\ \sigma_{33} \\ \tau_{12} \\ \tau_{13} \\ \tau_{23} \end{Bmatrix} \tag{3.1}$$

where $\frac{\nu_{ij}}{E_i} = \frac{\nu_{ji}}{E_j}$. The stiffness matrix is obtained by inverting the compliance matrix in Equation (3.1) (Rasheed 1996),

$$\begin{Bmatrix} \sigma_{11} \\ \sigma_{22} \\ \sigma_{33} \\ \tau_{12} \\ \tau_{13} \\ \tau_{23} \end{Bmatrix} = \begin{bmatrix} \frac{1-\nu_{23}\nu_{32}}{\Delta}E_1 & \frac{\nu_{21}+\nu_{31}\nu_{23}}{\Delta}E_1 & \frac{\nu_{31}+\nu_{21}\nu_{32}}{\Delta}E_1 & 0 & 0 & 0 \\ \frac{\nu_{21}+\nu_{31}\nu_{23}}{\Delta}E_1 & \frac{1-\nu_{13}\nu_{31}}{\Delta}E_2 & \frac{\nu_{32}+\nu_{12}\nu_{31}}{\Delta}E_2 & 0 & 0 & 0 \\ \frac{\nu_{31}+\nu_{21}\nu_{32}}{\Delta}E_1 & \frac{\nu_{32}+\nu_{12}\nu_{31}}{\Delta}E_2 & \frac{1-\nu_{21}\nu_{12}}{\Delta}E_3 & 0 & 0 & 0 \\ 0 & 0 & 0 & G_{12} & 0 & 0 \\ 0 & 0 & 0 & 0 & G_{13} & 0 \\ 0 & 0 & 0 & 0 & 0 & G_{23} \end{bmatrix} \begin{Bmatrix} \varepsilon_{11} \\ \varepsilon_{22} \\ \varepsilon_{33} \\ \gamma_{12} \\ \gamma_{13} \\ \gamma_{23} \end{Bmatrix} \tag{3.2}$$

where $\Delta = 1 - \nu_{12}\nu_{21} - \nu_{23}\nu_{32} - \nu_{13}\nu_{31} - 2\nu_{21}\nu_{32}\nu_{13}$.

If the compliance matrix in Equation (3.1) is reduced to 2-D behavior (sheet analysis), the stress components $\sigma_{33} = \tau_{13} = \tau_{23} = 0$. The third, fifth, and sixth rows and columns are removed, yielding

$$\begin{Bmatrix} \varepsilon_{11} \\ \varepsilon_{22} \\ \gamma_{12} \end{Bmatrix} = \begin{bmatrix} \frac{1}{E_1} & -\frac{\nu_{21}}{E_2} & 0 \\ -\frac{\nu_{12}}{E_1} & \frac{1}{E_2} & 0 \\ 0 & 0 & \frac{1}{G_{12}} \end{bmatrix} \begin{Bmatrix} \sigma_{11} \\ \sigma_{22} \\ \tau_{12} \end{Bmatrix} \tag{3.3}$$

The 2-D compliance matrix in Equation (3.3) may be inverted to yield the 2-D stiffness matrix (Jones 1975),

$$\begin{Bmatrix} \sigma_{11} \\ \sigma_{22} \\ \tau_{12} \end{Bmatrix} = \begin{bmatrix} \frac{E_1}{1-\nu_{12}\nu_{21}} & \frac{\nu_{12}E_2}{1-\nu_{12}\nu_{21}} & 0 \\ \frac{\nu_{21}E_1}{1-\nu_{12}\nu_{21}} & \frac{E_2}{1-\nu_{12}\nu_{21}} & 0 \\ 0 & 0 & G_{12} \end{bmatrix} \begin{Bmatrix} \varepsilon_{11} \\ \varepsilon_{22} \\ \gamma_{12} \end{Bmatrix} \tag{3.4}$$

3.6 FRP SHEET ENGINEERING CONSTANTS FROM CONSTITUENT PROPERTIES

Using the mechanics-of-materials approach requires certain simplifying assumptions in order to derive the mechanical properties of a unidirectional composite sheet. The accuracy of the estimated property depends on the accuracy of the assumption made.

3.6.1 Determination of E_1

The first modulus along the fiber direction may be determined by the rule of mixtures that results from the assumption of having the fiber and the matrix deform in equal amounts along the fiber direction (Jones 1975). This assumption is known to be very accurate, leading to an accurate estimation of the apparent Young's modulus E_1,

$$E_1 = E_f V_f + E_m V_m \tag{3.5}$$

where E_f is the fiber modulus, V_f is the fiber volume fraction, E_m is the matrix modulus, and $V_m = 1 - V_f$.

3.6.2 Determination of E_2

The second modulus along the transverse direction is not as straightforward to derive. One simplifying assumption can be made considering the same transverse stress σ_2 in the fiber and the matrix, leading to the following mechanics-of-materials expression, which is known to yield a lower bound value of the apparent Young's modulus E_2:

$$E_2 = \frac{E_f E_m}{V_m E_f + V_f E_m} \quad \text{or} \quad \frac{1}{E_2} = \frac{V_m}{E_m} + \frac{V_f}{E_f} \tag{3.6}$$

More accurate determination of E_2 could be obtained using the Halpin-Tsai equations (Jones 1975),

$$\frac{E_2}{E_m} = \frac{1 + \xi\eta V_f}{1 - \eta V_f} \tag{3.7}$$

where

$$\eta = \frac{E_f/E_m - 1}{E_f/E_m + \xi} \tag{3.8}$$

The value of ξ could be difficult to obtain, since it is a function of the fiber geometry, packing geometry, and loading conditions. However, studies have shown that a value of $\xi = 2$ can be approximated for calculating E_2 with a fiber volume fraction of 0.55 (Jones 1975). However, $\xi = 1$ has been observed to yield more accurate results when computing typical properties.

3.6.3 Determination of ν_{12}

The major Poisson's ratio ν_{12} may be determined by the rule of mixtures resulting from the previous two assumptions of having the fiber and the matrix deform in equal amounts along the fiber direction and having the transverse stress $\sigma_2 = 0$ (Jones 1975). These assumptions are known to be accurate, leading to an accurate estimation of the major Poisson's ratio ν_{12}:

$$\nu_{12} = \nu_f V_f + \nu_m V_m \tag{3.9}$$

3.6.4 Determination of G_{12}

The sheet in-plane shear modulus G_{12} is determined in the mechanics-of-materials approach using the assumption that the shearing stress of the fiber and the matrix are identical. The well-known nonlinear shear stress–strain is linearized using this assumption. Accordingly, the resulting equation yields a lower bound solution to the in-plane shear modulus G_{12}:

$$G_{12} = \frac{G_f G_m}{V_m G_f + V_f G_m} \quad \text{or} \quad \frac{1}{G_{12}} = \frac{V_m}{G_m} + \frac{V_f}{G_f} \tag{3.10}$$

$$G_f = \frac{E_f}{2(1+\nu_f)} \tag{3.11}$$

$$G_m = \frac{E_m}{2(1+\nu_m)} \tag{3.12}$$

More-accurate determination of G_{12} could be obtained using the Halpin-Tsai equations (Jones 1975),

$$\frac{G_{12}}{G_m} = \frac{1+\xi\eta V_f}{1-\eta V_f} \tag{3.13}$$

where

$$\eta = \frac{G_f/G_m - 1}{G_f/G_m + \xi} \tag{3.14}$$

The value of ξ could be difficult to obtain, since it is a function of the fiber geometry, packing geometry, and loading conditions. However, studies have shown that a value of $\xi = 1$ for calculating G_{12} can be approximated for a fiber volume fraction of 0.55 (Jones 1975).

3.6.5 Determination of ν_{21}:

Once the first three parameters are estimated, the minor Poisson's ratio ν_{21} is directly calculated, as discussed in Section 3.5:

$$\frac{\nu_{12}}{E_1} = \frac{\nu_{21}}{E_2} \Rightarrow \nu_{21} = \frac{\nu_{12}}{E_1} E_2 \tag{3.15}$$

Since the ratio of $\frac{\nu_{12}}{E_1}$ is accurately estimated, the minor Poisson's ratio ν_{21} will yield a lower bound solution if the mechanics-of-materials approach is followed.

Example 3.1

E-Glass FRP composite is made of 60% fiber and 40% epoxy matrix by volume. Determine the in-plane orthotropic properties of the composite sheet from the constituent properties.

$\nu_f = 0.22$

$\nu_m = 0.37$

Solution:

Using Tables 3.2 and 3.3,

$$E_1 = 72.4 \times 0.6 + 4 \times 0.4 = 45.04 \text{ GPa}$$

$$E_2 = \frac{72.4 \times 4}{0.4 \times 72.4 + 0.6 \times 4} = 9.235 \text{ GPa} \quad \text{(mechanics of materials)}$$

$$\eta = \frac{72.4/4 - 1}{72.4/4 + 1} = 0.895$$

$$E_2 = \frac{1 + 1 \times 0.895 \times 0.6}{1 - 0.895 \times 0.6} \times 4 = 13.28 \text{ GPa} \quad \text{(Halpin-Tsai)}$$

$$\nu_{12} = 0.22 \times 0.6 + 0.37 \times 0.4 = 0.28$$

$$\nu_{21} = 0.057 \quad \text{(mechanics of materials)}$$

$$\nu_{21} = 0.0826 \quad \text{(Halpin-Tsai)}$$

$$G_f = \frac{72.4}{2(1.22)} = 29.67 \text{ GPa}$$

$$G_m = \frac{4}{2(1.37)} = 1.46 \text{ GPa}$$

$$G_{12} = \frac{29.67 \times 1.46}{0.4 \times 29.67 + 0.6 \times 1.46} = 3.4 \text{ GPa} \quad \text{(mechanics of materials)}$$

$$\eta = \frac{29.67/1.46 - 1}{29.67/1.46 + 1} = 0.906$$

$$G_{12} = \frac{1 + 1 \times 0.906 \times 0.6}{1 - 0.906 \times 0.6} \times 1.46 = 4.94 \text{ GPa} \quad \text{(Halpin-Tsai)}$$

Compare to typical numbers of E-GFRP epoxy with 60% fiber (Table3.7)

$$E_1 = 45 \text{ GPa}$$

$$E_2 = 12 \text{ GPa}$$

$$\nu_{12} = 0.28$$

$$\nu_{21} = 0.075$$

$$G_{12} = 5.5 \text{ GPa}$$

3.7 PROPERTIES OF FRP COMPOSITES (TENSION)

FRP composites provide stiffness and strength along the fiber direction in tension, and they behave linearly elastic along that direction up to brittle material failure or rupture, as shown in Figure 3.9. On the other hand, the behavior along the matrix or transverse direction and in-plane shear is generally nonlinear. The behavior along the transverse direction could be approximated as linear elastic up to matrix

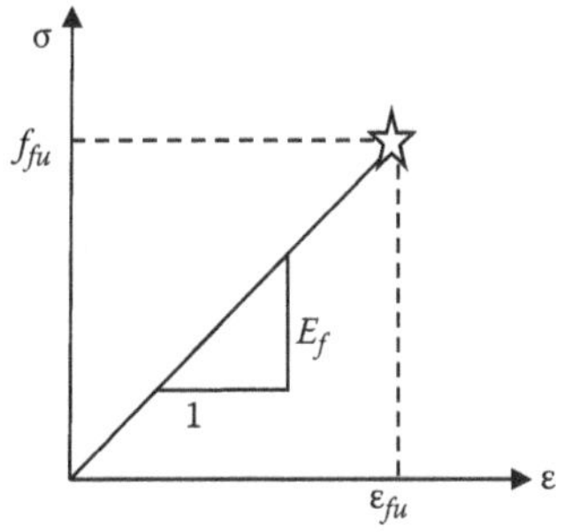

FIGURE 3.9 FRP design stress–strain curve along the fiber direction.

cracking, while in-plane shear behavior is always nonlinear. Nevertheless, this text focuses on the behavior along the fiber direction, since FRP fibers are aligned with the structural axis of loading in the strengthening application.

According to ACI 440.2R-08,

$$\overline{f}_{fu} \equiv \text{ mean ultimate tensile strength of 20 coupon specimens}$$

$$\overline{\varepsilon}_{fu} \equiv \text{ mean ultimate tensile strength of 20 coupon specimens}$$

based on ASTM D3039 and D7205.

$$f^*_{fu} = \overline{f}_{fu} - 3\sigma_f = \text{ ultimate tensile strength} \tag{3.16}$$

$$\varepsilon^*_{fu} = \overline{\varepsilon}_{fu} - 3\sigma_\varepsilon = \text{ ultimate tensile strain} \tag{3.17}$$

where σ_f and σ_ε are the standard deviations for the ultimate strength and strain according to ACI 440.2R-08.

Equations (3.16) and (3.17) yield 99.87% probability of exceedance

$$f_{fu} = C_E f^*_{fu} = \text{ design ultimate tensile strength} \tag{3.18}$$

$$\varepsilon_{fu} = C_E \varepsilon^*_{fu} = \text{ design ultimate tensile strain} \tag{3.19}$$

$$E_f = \frac{f_{fu}}{\varepsilon_{fu}} = \frac{f^*_{fu}}{\varepsilon^*_{fu}} = \text{ design modulus of elasticity of FRP} \tag{3.20}$$

where C_E is the environmental reduction coefficient (ACI 440.2R-08), as seen in Table 3.5.

For FRP plates or precured laminates, properties that are reported are those of the composite, since fiber and resin volume fractions are well controlled. Conversely, FRP sheets manufactured by hand wet lay-up have their properties primarily reported based on the net fiber area, since it is more controlled than the fixed width

TABLE 3.5
Environmental Reduction Factors for Different Exposure Cases and FRP Systems

Exposure Conditions	Fiber Type	Environmental Reduction Factor, C_E
Interior exposure	Carbon	0.95
	Glass	0.75
	Aramid	0.85
Exterior exposure (bridges, piers, and unenclosed parking garages)	Carbon	0.85
	Glass	0.65
	Aramid	0.75
Aggressive environment (chemical plants and wastewater treatment plants)	Carbon	0.85
	Glass	0.50
	Aramid	0.70

Source: Courtesy of ACI 440.2R-08.

and thickness of the cured composite system. This is due to the fact that the wet lay-up process has controlled fiber content and variable resin content that depends on the installer (ACI 440.2R-08). It is very important to note that net fiber area properties are *not* the same as dry fiber properties, but rather the same as the laminate or composite properties with known fiber content. Since the thickness of the FRP sheet is very small, the corresponding strength and modulus values are high because ($f_{fu}A_f$) or (E_fA_f) are constant for a composite.

For practical and design purposes, FRP mechanical properties are needed. ACI 440R-07 reports typical values for sheets and plates from various manufacturers for materials used in civil infrastructure-strengthening applications, as shown in Table 3.6. It can be seen from the values in Table 3.6 that CFRP typically has higher modulus and tensile strength and lower ultimate strain than GFRP, as shown in Figure 3.10.

On the other hand, the *Delaware Encyclopedia of Composites* (Zweben 1989) reports typical composite mechanical properties used in aerospace applications. These values are listed here for comparison purposes. It is evident that the laminate properties of the materials used for aerospace applications clearly exceed those of plates used for civil engineering applications due to the higher control over the manufacturing process involved (heating, pressure, and vacuum). On the other hand, the FRP sheet properties (Table 3.6) may be seen to exceed those of aerospace composites (Tables 3.7–3.12), which is due to the fact that sheet properties are based on net fiber area and *not* on composite laminate area. For example, MBrace EG 900 glass sheet has a net fiber modulus of 72.4 GPa, and multiplying this by the fiber volume fraction of 0.6 yields 43.44 GPa for the composite, which is close to the 45 GPa reported in Table 3.7 for E-glass. Similarly, the tensile strength has a net fiber value of 1517 MPa, and multiplying this by 0.6 yields 910 MPa for the composite, which is slightly less than the 1020 MPa reported in Table 3.7. Further properties of unidirectional, cross-ply and angle-ply laminates are given in Table 3.13 (ACI 440.2R-08).

TABLE 3.6
Manufacturer Mechanical Properties of Some Commercially Available Systems

FRP System	Fiber Type	Weight, g/m^2 (lb/ft^2)	Design Thickness, mm (in.)	Tensile Strength, MPa (ksi)	Tensile Elastic Modulus, GPa (ksi)	ACI 440.3R Test Reporting Method
			Fyfe Co. LLC (2005)			
Tyfo SEH51 sheet	Glass	915 (0.19)	1.3 (0.052)	575 (83.4)	26.1 (3,785)	Method 1
Tyfo SCH41 sheet	Carbon	644 (0.14)	1.0 (0.040)	985 (143)	95.8 (13,900)	Method 1
			Sika Corp. (2007)			
SikaWrap Hex 100G sheet	Glass	913 (0.19)	1.0 (0.040)	531 (77)	23.6 (3,430)	Method 1
SikaWrap Hex 103C sheet	Carbon	618 (0.13)	1.0 (0.040)	717 (104)	65.1 (9,450)	Method 1
CarboDur S plate	Carbon	1800 (0.37)	1.2–1.4 (0.048–0.055)	2800 (406)	165 (23,900)	Method 1
CarboDur M plate	Carbon	1900 (0.39)	1.2 (0.048)	2400 (348)	210 (30,500)	Method 1
CarboDur H plate	Carbon	1900 (0.39)	1.2 (0.048)	1300 (189)	300 (43,500)	Method 1
			BASF (2006)			
MBrace EG 900 sheet	Glass	900 (0.19)	0.37 (0.015)	1517 (220)	72.4 (10,500)	Method 2
MBrace AK 60 sheet	Aramid	600 (0.12)	0.28 (0.011)	2000 (290)	120 (17,400)	Method 2
MBrace CF 130	Carbon	300 (0.062)	0.17 (0.007)	3800 (550)	227 (33,000)	Method 2
MBrace CF 160	Carbon	600 (0.124)	0.33 (0.013)	3800 (550)	227 (33,000)	Method 2
S&P 100/1.4	Carbon	—	1.4 (0.055)	2700 (390)	159 (23,000)	Method 1
			Hughes Brothers (2005)			
Aslan 400 plate	Carbon	—	1.4 (0.055)	2400 (350)	131 (19,000)	Method 1
Aslan 500 tape	Carbon	—	2.0 (0.079)	2068 (300)	124 (18,000)	Method 1
Aslan 500 tape	Carbon	—	4.5 (0.177)	1965 (285)	124 (18,000)	Method 1

Source: Courtesy of ACI 440R-07.

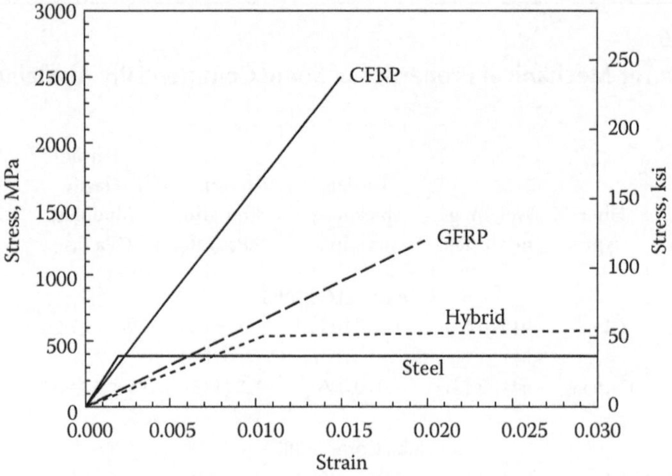

FIGURE 3.10 Typical stress–strain curves for FRP systems. (Courtesy of ACI 440R-07.)

TABLE 3.7
Typical Mechanical Properties for E-Glass/Epoxy Unidirectional Composites with $V_f = 0.6$

Elastic Constants	**GPa**	**10^6 psi**
Longitudinal modulus, E_L	45	6.5
Transverse modulus, E_T	12	1.8
Axial shear modulus, G_{LT}	5.5	0.8
Poisson's ratio, ν_{LT} (dimensionless)		0.28
Strength Properties	**MPa**	**10^3 psi**
Longitudinal tension, F_L^{tu}	1020	150
Longitudinal compression, F_L^{cu}	620	90
Transverse tension, F_T^{tu}	40	7
Transverse compression, F_T^{cu}	140	20
In-plane shear, F_{LT}^{su}	70	10
Interlaminar shear, F^{isu}	70	10
Ultimate Strains	**%**	
Longitudinal tension, ϵ_L^{tu}	2.3	
Longitudinal compression, ϵ_L^{cu}	1.4	
Transverse tension, ϵ_T^{tu}	0.4	
Transverse compression, ϵ_T^{cu}	1.1	
In-plane shear, γ_{LT}^{u}	1–6	
Density, kg/m^3 (lb/in.^{3})	2.1×10^3 (0.075)	

Source: Courtesy of Zweben (1989).

TABLE 3.8
Typical Mechanical Properties for S-Glass/Epoxy Unidirectional Composites with $V_f = 0.6$

Elastic Constants	**GPa**	**10^6 psi**
Longitudinal modulus, E_L	55	8.0
Transverse modulus, E_T	16	2.3
Axial shear modulus, G_{LT}	7.6	1.1
Poisson's ratio, ν_{LT} (dimensionless)		0.28
Strength Properties	**MPa**	**10^3 psi**
Longitudinal tension, F_L^{tu}	1620	230
Longitudinal compression, F_L^{cu}	690	100
Transverse tension, F_T^{tu}	40	7
Transverse compression, F_T^{cu}	140	20
In-plane shear, F_{LT}^{su}	80	12
Interlaminar shear, F^{isu}	80	12
Ultimate Strains	**%**	
Longitudinal tension, ϵ_L^{tu}	2.9	
Longitudinal compression, ϵ_L^{cu}	1.3	
Transverse tension, ϵ_T^{tu}	0.4	
Transverse compression, ϵ_T^{cu}	1.1	
In-plane shear, γ_{LT}^{u}	1–6	
Density, kg/m³ (lb/in.³)	2.0×10^3 (0.073)	

Source: Courtesy of Zweben (1989).

Example 3.2

Two test panels were fabricated, one using one ply only and the other one using two plies of an E-glass fiber/epoxy unidirectional FRP system with the wet lay-up technique (Wuertz 2013). Based on the manufacturer's data sheet of this FRP system, the net fiber area is 0.043 in.^{2}/in. (1.092 mm^2/mm) width per ply. After the system has cured, three 1-in. (25.4 mm)-wide test coupons are cut from the single-ply panel (GFRP-1, GFRP-2, GFRP-3), and three 1-in. (25.4 mm)-wide test coupons are cut from the two-ply panel (GFRP-4, GFRP-5, GFRP-6). The test coupons are tested in tension to failure according to ASTM D3039. Table 3.14 presents the results of the tension tests. Determine the average tensile properties based on the composite laminate and the net fiber area.

Solution:

Properties for the composite laminate:

TABLE 3.9
Typical Mechanical Properties for High-Strength Graphite/Epoxy Unidirectional Composites with $V_f = 0.6$

Elastic Constants	**GPa**	**10^6 psi**
Longitudinal modulus, E_L	145	21
Transverse modulus, E_T	10	1.5
Axial shear modulus, G_{LT}	4.8	0.7
Poisson's ratio, ν_{LT} (dimensionless)		0.25
Strength Properties	**MPa**	**10^3 psi**
Longitudinal tension, F_L^{tu}	1240	180
Longitudinal compression, F_L^{cu}	1240	180
Transverse tension, F_T^{tu}	41	6
Transverse compression, F_T^{cu}	170	25
In-plane shear, F_{LT}^{su}	80	12
Interlaminar shear, F^{isu}	80	12
Ultimate Strains	**%**	
Longitudinal tension, ϵ_L^{tu}	0.9	
Longitudinal compression, ϵ_L^{cu}	0.9	
Transverse tension, ϵ_T^{tu}	0.4	
Transverse compression, ϵ_T^{cu}	1.6	
In-plane shear, γ_{LT}^{u}	1–6	
Density, kg/m³ (lb/in.³)	1.58×10^3 (0.057)	

Source: Courtesy of Zweben (1989).

For the first three coupons,

$$A_f = 0.104 \times 1 = 0.104 in^2 (67.1 mm^2)$$

$$\bar{f}_{fu} = \frac{3.382}{0.104} = 32.52 ksi (224.3\ MPa)$$

$$E_f = \frac{\bar{f}_{fu}}{\varepsilon_{fu}} = \frac{32.52}{0.0162} = 2007.4\ ksi\ (13.85 GPa)$$

Compared to 2010 ksi (13.867 GPa) measured at 60% of the rupture load. For the other three coupons,

$$A_f = 0.153 \times 1 = 0.153 in^2 (98.71 mm^2)$$

$$\bar{f}_{fu} = \frac{6.828}{0.153} = 44.63 ksi (307.9\ MPa)$$

$$E_f = \frac{\bar{f}_{fu}}{\varepsilon_{fu}} = \frac{44.63}{0.0192} = 2324.4 ksi (16.04\ GPa)$$

Compared to 2326.3 ksi (16.05 GPa) measured at 60% of the rupture load.

TABLE 3.10
Typical Mechanical Properties for High-Modulus Graphite/ Epoxy Unidirectional Composites with $V_f = 0.6$

Elastic Constants	GPa	10^6 psi
Longitudinal modulus, E_L	220	32
Transverse modulus, E_T	6.9	1.0
Axial shear modulus, G_{LT}	4.8	0.7
Poisson's ratio, v_{LT} (dimensionless)		0.25
Strength Properties	**MPa**	**10^3 psi**
Longitudinal tension, F_L^{tu}	760	110
Longitudinal compression, F_L^{cu}	690	100
Transverse tension, F_T^{tu}	28	4
Transverse compression, F_T^{cu}	170	25
In-plane shear, F_{LT}^{su}	70	10
Interlaminar shear, F^{isu}	70	10
Ultimate Strains	**%**	
Longitudinal tension, ϵ_L^{tu}	0.3	
Longitudinal compression, ϵ_L^{cu}	0.3	
Transverse tension, ϵ_T^{tu}	0.4	
Transverse compression, ϵ_T^{cu}	2.8	
In-plane shear, $\gamma^u{}_{LT}$	1–6	
Density, kg/m^3 (lb/in.^{3})	1.64×10^3 (0.059)	

Source: Courtesy of Zweben (1989).

It is evident that the two sets of coupons have drastically different strengths, even though they are made of the same materials. This is due to the fact that the second set (with two plies) has a higher fiber volume fraction. To overcome this difficulty, the net fiber area is used.

Properties for the net fiber area:

For the first three coupons,

$$A_f = 0.043\text{in}^2 / in \times 1in = 0.043\text{in}^2 (27.74\ mm^2)$$

$$\bar{f}_{fu} = \frac{3.382}{0.043} = 78.65\text{k}si\ (542.6\ MPa)$$

$$E_f = \frac{\bar{f}_{fu}}{\bar{\varepsilon}_{fu}} = \frac{78.65}{0.022} = 3575\text{k}si\ (24.66\ GPa)$$

TABLE 3.11
Typical Mechanical Properties for Ultrahigh-Modulus Graphite/Epoxy Unidirectional Composites with $V_f = 0.6$

Elastic Constants	GPa	10^6 psi
Longitudinal modulus, E_L	290	42
Transverse modulus, E_T	6.2	0.9
Axial shear modulus, G_{LT}	4.8	0.7
Poisson's ratio, ν_{LT} (dimensionless)		0.25
Strength Properties	**MPa**	**10^3 psi**
Longitudinal tension, F_L^{tu}	620	90
Longitudinal compression, F_L^{cu}	620	90
Transverse tension, F_T^{tu}	21	3
Transverse compression, F_T^{cu}	170	25
In-plane shear, F_{LT}^{su}	60	9
Interlaminar shear, F^{isu}	60	9
Ultimate Strains	**(%)**	
Longitudinal tension, ϵ_L^{tu}	0.2	
Longitudinal compression, ϵ_L^{cu}	0.2	
Transverse tension, ϵ_T^{tu}	0.3	
Transverse compression, ϵ_T^{cu}	2.8	
In-plane shear, γ_{LT}^{u}	0.6–4	
Density, kg/m³ (lb/in.³)	1.70×10^3 (0.061)	

Source: Courtesy of Zweben (1989).

The ultimate strain here is reported by the manufacturer.

For the other three coupons,

$$A_f = 2 \times 0.043 \times 1 = 0.086\text{in}^2 \,(55.48\ mm^2)$$

$$\bar{f}_{fu} = \frac{6.828}{0.086} = 79.4ksi\ (547.75\ MPa)$$

$$E_f = \frac{\bar{f}_{fu}}{\bar{\varepsilon}_{fu}} = \frac{79.4}{0.022} = 3609ksi\ (24.9\ GPa)$$

The ultimate strain here is reported by the manufacturer.

It is evident that the two sets of coupons have almost the same net fiber strength and modulus. This is why net fiber properties are more consistent to use in strengthening than laminate properties for wet lay-up applications. Another way to obtain consistent results between the two sets of coupons is to compute the average FRP strength per unit width of the laminate:

TABLE 3.12
Typical Mechanical Properties for Kevlar 49 Aramid/Epoxy Unidirectional Composites with $V_f = 0.6$

Elastic Constants	**GPa**	**10^6 psi**
Longitudinal modulus, E_L	76	11
Transverse modulus, E_T	5.5	0.8
Axial shear modulus, G_{LT}	2.1	0.3
Poisson's ratio, ν_{LT} (dimensionless)		0.34
Strength Properties	**MPa**	**10^3 psi**
Longitudinal tension, F_L^{tu}	1240	180
Longitudinal compression, "yield," F_L^{cy}	230	33
Longitudinal compression, ultimate, F_L^{cu}	280	40
Transverse tension, F_T^{tu}	30	4.3
Transverse compression, F_T^{cu}	140	20
In-plane shear, F_{LT}^{su}	60	9
Interlaminar shear, F^{isu}	60	9
Ultimate Strains	**%**	
Longitudinal tension, ϵ_L^{tu}	1.6	
Longitudinal compression, "yield," ϵ_L^{cy}	0.3	
Longitudinal compression, ultimate, ϵ_L^{cu}	>2.0	
Transverse tension, ϵ_T^{tu}	0.5	
Transverse compression, ϵ_T^{cu}	2.5	
In-plane shear, γ_{LT}^{u}	1–6	
Density, kg/m³ (lb/in.³)	1.38×10^3 (0.050)	

Source: Courtesy of Zweben (1989).

For the first three coupons:

$$A_f = 0.104 \times 1 = 0.104 \text{in}^2 (67.1\, mm^2)$$

$$\bar{p}_{fu} = \frac{\bar{f}_{fu} A_f}{n w_f} = \frac{32.52 \times 0.104}{1 \times 1} = 3.382 \text{k} / in / ply\ (0.593\ kN / mm / ply)$$

For the other three coupons:

$$A_f = 0.153 \times 1 = 0.153 \text{in}^2 (98.71\, mm^2)$$

$$\bar{p}_{fu} = \frac{\bar{f}_{fu} A_f}{n w_f} = \frac{44.63 \times 0.153}{2 \times 1} = 3.414 \text{k} / in / ply\ (0.598\ kN / mm / ply)$$

TABLE 3.13
Range of Typical Tensile Properties for FRP Considering Three Laminate Lay-Ups with Fiber Volumes of 40%–60%

FRP System Description (Fiber Orientation)[a]	Young's Modulus		Ultimate Tensile Strength		Rupture Strain at 0°, %
	Property at 0°, 10^3 ksi (GPa)	Property at 90°, 10^3 ksi (GPa)	Property at 0°, ksi (MPa)	Property at 90°, ksi (MPa)	
High-Strength Carbon/Epoxy					
0°	15 to 21 (100 to 140)	0.3 to 1 (2 to 7)	150 to 350 (1020 to 2080)	5 to 10 (35 to 70)	1.0 to 1.5
0°/90°	8 to 11 (55 to 76)	8 to 11 (55 to 75)	100 to 150 (700 to 1020)	100 to 150 (700 to 1020)	1.0 to 1.5
+45°/–45°	2 to 4 (14 to 28)	2 to 4 (14 to 28)	25 to 40 (180 to 280)	25 to 40 (180 to 280)	1.5 to 2.5
E-Glass/Epoxy					
0°	3 to 6 (20 to 40)	0.3 to 1 (2 to 7)	75 to 200 (520 to 1400)	5 to 10 (35 to 70)	1.5 to 3.0
0°/90°	2 to 5 (14 to 34)	2 to 5 (14 to 35)	75 to 150 (520 to 1020)	75 to 150 (520 to 1020)	2.0 to 3.0
+45°/–45°	2 to 3 (14 to 21)	2 to 3 (14 to 20)	25 to 40 (180 to 280)	25 to 40 (180 to 280)	2.5 to 3.5
High-Performance Aramid/Epoxy					
0°	7 to 10 (48 to 68)	0.3 to 1 (2 to 7)	100 to 250 (700 to 1720)	5 to 10 (35 to 70)	2.0 to 3.0
0°/90°	4 to 5 (28 to 34)	4 to 5 (28 to 34)	40 to 80 (280 to 550)	40 to 80 (280 to 550)	2.0 to 3.0
+45°/–45°	1 to 2 (7 to 14)	1 to 2 (7 to 14)	20 to 30 (140 to 210)	20 to 30 (140 to 210)	2.0 to 3.0

Source: Courtesy of ACI 440.2R-08.

Notes: FRP composite properties are based on FRP systems having an approximate fiber volume of 50% and a composite thickness of 0.1 in. (2.5 mm). In general, FRP bars have fiber volumes of 50%–70%. Precured systems have fiber volumes of 40%–60%, and wet lay-up systems have fiber volumes of 25%–40%. Because the fiber volume influences the gross laminate properties, precured laminates usually have higher mechanical properties than laminates created using the wet lay-up technique.

[a] Zero degrees represents unidirectional fiber orientation. Zero/90° or +45°/–45° represent fiber balanced in two orthogonal directions, where 0° is the direction of loading, and 90° is normal to the direction of loading. Tension is applied to 0° direction. All FRP bar properties are in the 0° direction.

TABLE 3.14
GFRP Coupon Testing To Determine Laminate and Net-Fiber Properties

Specimen	Width (in.)	Width (mm)	Average Thickness (in.)	Average Thickness (mm)	Rupture Load (kips)	Rupture Load (kN)	Ultimate Strain
GFRP-1	1.00	25.4	0.099	2.51	3.237	14.405	0.01507
GFRP-2	1.00	25.4	0.103	2.62	3.368	14.988	0.0166
GFRP-3	1.00	25.4	0.111	2.82	3.541	15.757	0.01694
Average	**1.00**	**25.4**	**0.104**	**2.65**	**3.382**	**15.05**	**0.0162**
GFRP-4	1.00	25.4	0.166	4.22	7.387	32.872	0.02195
GFRP-5	1.00	25.4	0.142	3.61	5.467	24.328	0.01699
GFRP-6	1.00	25.4	0.152	3.86	7.630	33.955	0.01868
Average	**1.00**	**25.4**	**0.153**	**3.89**	**6.828**	**30.385**	**0.0192**

3.8 PROPERTIES OF FRP COMPOSITES (COMPRESSION)

ACI 440.2R-08 does not allow externally bonded FRP sheets or plates to carry compression due to the lack of testing or experimental validation. Nevertheless, Wu (1990) reported that the compressive strength of coupons is 55% of the tensile strength for GFRP, 78% for CFRP, and 20% for AFRP, depending on the fiber volume fraction and the resin used. Composites with fiber volume fraction of 55%–60% are reported to have a compressive modulus equal to 80% of the tensile modulus for GFRP, 85% for CFRP, and 100% for AFRP (Ehsani 1993). These results are reported primarily for FRP bars, but not for sheets or plates externally bonded to concrete using resin adhesives.

3.9 PROPERTIES OF FRP COMPOSITES (DENSITY)

Composites are known to be very lightweight materials, ranging in weight between 15%–27% of the weight of steel, as seen in Table 3.15. Accordingly, significant savings in transportation costs, dead-load reduction, and on-site ease of installation are anticipated (ACI 440.2R-08).

3.10 PROPERTIES OF FRP COMPOSITES (THERMAL EXPANSION)

The coefficient of thermal expansion for unidirectional FRP systems varies significantly between the fiber and the transverse directions, based on the fiber and resin type as well as the fiber volume fraction. Typical values are listed in Table 3.16, where concrete and steel values are also listed for reference purposes.

To determine the values of the coefficients of thermal expansion (CTE) along the structural principal directions when the material principal directions are different (i.e., in the case of angle plies of $\pm\theta°$ between the fiber and the main structural loading axis), the generalized strain–stress relationship, in the presence of temperature changes, is invoked:

TABLE 3.15
Typical Densities of FRP Materials Compared to Steel

Material	Density lb/ft³ (g/cm³)
Steel	490 (7.9)
GFRP	75–130 (1.2–2.1)
CFRP	90–100 (1.5–1.6)
AFRP	75–90 (1.2–1.5)

Along the material principal axes:

$$\begin{Bmatrix} \varepsilon_1 \\ \varepsilon_2 \\ \gamma_{12} \end{Bmatrix} = \begin{bmatrix} C_{11} & C_{12} & 0 \\ C_{12} & C_{22} & 0 \\ 0 & 0 & C_{66} \end{bmatrix} \begin{Bmatrix} \sigma_1 \\ \sigma_2 \\ \tau_{12} \end{Bmatrix} + \begin{Bmatrix} \alpha_1 \\ \alpha_2 \\ 0 \end{Bmatrix} \Delta T \tag{3.21}$$

Along the structure principal axes:

$$\begin{Bmatrix} \varepsilon_x \\ \varepsilon_y \\ \gamma_{xy} \end{Bmatrix} = \begin{bmatrix} \bar{C}_{11} & \bar{C}_{12} & \bar{C}_{16} \\ \bar{C}_{12} & \bar{C}_{22} & \bar{C}_{26} \\ \bar{C}_{16} & \bar{C}_{26} & \bar{C}_{66} \end{bmatrix} \begin{Bmatrix} \sigma_x \\ \sigma_y \\ \tau_{xy} \end{Bmatrix} + \begin{Bmatrix} \alpha_x \\ \alpha_y \\ \alpha_{xy} \end{Bmatrix} \Delta T \tag{3.22}$$

where the CTEs transform like tensor strains (see Figure 3.11),

$$\begin{Bmatrix} \alpha_x \\ \alpha_y \\ \dfrac{\alpha_{xy}}{2} \end{Bmatrix} = \begin{bmatrix} C^2 & S^2 & -2CS \\ S^2 & C^2 & +2CS \\ +CS & -CS & C^2 - S^2 \end{bmatrix} \begin{Bmatrix} \alpha_1 \\ \alpha_2 \\ 0 \end{Bmatrix} \tag{3.23}$$

TABLE 3.16
FRP Coefficients of Thermal Expansion vs. Steel and Concrete, × 10⁻⁶/°F (× 10⁻⁶/°C)

Direction	GFRP	CFRP	AFRP	Concrete	Steel
Longitudinal α_L	3.3 to 5.6 (6 to 10)	−0.6 to 0 (−1 to 0)	−3.3 to −1.1 (−6 to −2)	5.5 (9.9) [a]	6.5 (11.7) [a]
Transverse α_T	10.4 to 12.6 (19 to 23)	12 to 27 (22 to 50)	33 to 44 (60 to 80)	5.5 (9.9)	6.5 (11.7)

[a] From Beer and Johnston (1992).

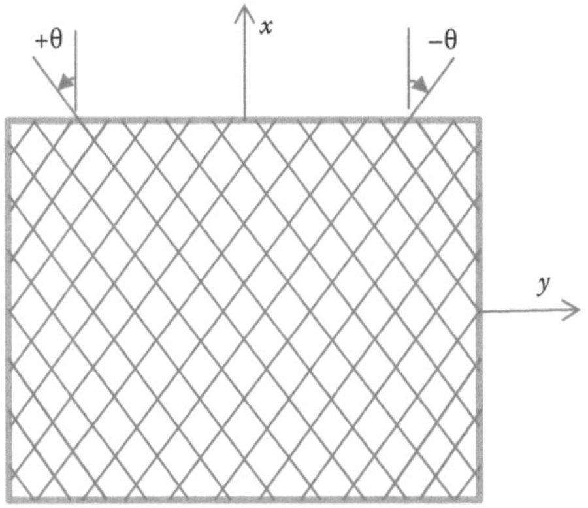

FIGURE 3.11 Typical angle ply laminate with fiber orientations with respect to the loading axes.

Along the angle-ply laminate:
Combining the CTEs of the two identical plies together,

$$\alpha_x = 0.5\left(\alpha_1 C^2 + \alpha_2 S^2\right) + 0.5\left(\alpha_1 C^2 + \alpha_2 S^2\right) = \alpha_1 C^2 + \alpha_2 S^2$$

$$\alpha_y = 0.5\left(\alpha_1 S^2 + \alpha_2 C^2\right) + 0.5\left(\alpha_1 S^2 + \alpha_2 C^2\right) = \alpha_1 S^2 + \alpha_2 C^2 \qquad (3.24)$$

$$\frac{\alpha_{xy}}{2} = 0.5(\alpha_1 CS - \alpha_2 CS) + 0.5(-\alpha_1 CS + \alpha_2 CS) = 0$$

3.11 PROPERTIES OF FRP COMPOSITES (HIGH TEMPERATURE)

The glass transition temperature, T_g, is the threshold beyond which the polymer softens and loses its ability to transfer stress from concrete to fiber. Typical T_g values for resins cured at room temperature range between 140°F–180°F (60°C–82°C). In bond-critical applications, reaching T_g would be the cutoff point for the FRP stress-carrying capacity. In contact-critical applications, reduced stress along the fiber may be carried up to 1800°F (1000°C) for CFRP, 350°F (175°C) for AFRP, and 530°F (275°C) for GFRP (ACI 440.2R-08). Reduction in the tensile strength of CFRP exceeds 20% at 500°F (260°C), according to Hawileh et al. (2009).

3.12 PROPERTIES OF FRP COMPOSITES (LONG-TERM EFFECTS)

With FRP composites, long-term effects primarily include creep rupture and fatigue. Creep rupture is a sudden failure at a sustained tensile loading for an extended period of time. The time-to-failure period is reduced when the sustained tensile stress to short-term ultimate strength ratio is increased and when the environmental exposure is increased (ACI 440.2R-08).

Two studies on FRP bars concluded that maintaining the sustained stress ratio below 0.3 for GFRP, 0.5 for AFRP, and 0.9 for CFRP results in time-to-failure periods exceeding 500,000 hours (about 57 years). These studies were conducted by Yamaguchi et al. (1997) and Malvar (1998).

Fatigue is also a sudden failure at a cyclic relatively low stress range for an extended period of time. For a stress ratio (minimum to maximum applied stress) of 0.1 using tension-tension sinusoidal low-cycle loading, the fatigue strength of CFRP is 60%–70% of the static ultimate strength after 1 million cycles (ACI 440.2R-08). Larson, Peterman, and Rasheed (2005) performed experiments on full-scale concrete beams prestressed with straight strands that were then strengthened with CFRP. The failure mode observed was CFRP rupture for all beams tested under static and fatigue loading. Furthermore, there was no reduction in the CFRP tensile strength at static failure after 1 million cycles of fatigue at about 124 MPa (18 ksi) strand stress range. However, the CFRP fatigue strength was about 79% of the static ultimate strength after 3 million cycles of fatigue at about 248 MPa (36 ksi) strand stress range (Rasheed, Larson, and Peterman 2006). On the other hand, GFRP undergoes 10% loss of static ultimate strength per decade of logarithmic life time (Mandell 1982). Odagiri, Matsumoto, and Nakai (1997) reported a tension-tension fatigue strength of AFRP in the range of 54%–73% of the static ultimate strength after 2 million cycles of fatigue. Odagiri, Matsumoto, and Nakai (1997) suggested limiting the maximum stress to 0.54–0.73 times the static tensile strength to avoid lifetime fatigue failure. It is important to note that all these figures apply to unidirectional composite with loading along the fiber direction.

Chapter Problems

Problem 3.1

Determine the laminate in-plane stiffness properties of AS-4 carbon/epoxy and Kevlar 49/epoxy for 60% fiber volume fraction. Then compare your results with the typical values given in the tables of the *Delaware Encyclopedia of Composites*. Use imperial units. The Possion's ratio for epoxy = 0.37, for AS-4 carbon fiber = 0.17, and for Kevlar 49 fiber = 0.32.

Problem 3.2

Determine the laminate in-plane stiffness properties of S-glass/epoxy for 60% fiber volume fraction. Then compare your results with the typical values given in the tables of the *Delaware Encyclopedia of Composites*. Use SI units. The Possion's ratio for epoxy = 0.37 and for S-glass fiber = 0.22.

Problem 3.3

A test panel was fabricated using one ply only of carbon fiber/epoxy unidirectional FRP system with the wet lay-up technique (Decker 2007). Based on the manufacturer's data sheet of this FRP system, the net fiber area is 0.0065 in.^{2}/in. (0.165 mm^2/mm) width per ply. After the system has cured, five 2-in. (50.8 mm)-wide test coupons are cut from the single-ply panel (CFRP-1 through CFRP-5). The test coupons were tested in tension to rupture according to ASTM D3039. The following table presents the results of the tension tests. Determine the average and the design tensile properties based on the composite laminate and the net fiber area.

Specimen	Width (in.)	Width (mm)	Average Thickness (in.)	Average Thickness (mm)	Rupture Load (kips)	Rupture Load (kN)	Ultimate Strain (με)
CFRP-1	2.00	50.8	0.0279	0.709	6.13	27.28	12,514
CFRP-2	2.00	50.8	0.0285	0.724	6.14	27.32	14,123
CFRP-3	2.00	50.8	0.0276	0.701	6.16	27.41	12,931
CFRP-4	2.00	50.8	0.0273	0.693	6.18	27.50	11,947
CFRP-5	2.00	50.8	0.0265	0.673	6.04	26.88	12,993
Average	2.00	50.8	0.0276	0.700	6.13	27.28	12,902

Problem 3.4

Using the density values in Table 3.15, determine the ratio of the tensile strength to density and tensile modulus to density per inch thickness (per mm thickness) and rank the four materials based on the two ratios from highest to lowest. Use Table 3.7 for GFRP, Table 3.9 for CFRP, and Table 3.12 for AFRP. Assume the tensile strength and modulus of steel to be 70 ksi and 29,000 ksi (483 MPa and 200 GPa), respectively.

Problem 3.5

Determine the angle–ply orientation needed to make the following FRP materials thermally compatible with concrete in the longitudinal direction.

a. GFRP
b. CFRP
c. AFRP

Take the average α_1 and α_2 for each from Table 3.16 and take $\alpha_{concrete} = 5.5 \times 10^{-6}/°F$. Indicate which composite of the three is the most thermally compatible with reinforced concrete in the longitudinal direction when used as a 0° unidirectional ply.

Problem 3.6

Design a $\pm\theta$ angle–ply laminate for each FRP composite indicated in Problems 3.5 to be thermally compatible with concrete in the transverse direction to limit the possibility of matrix cracking.

REFERENCES

ACI 440R-07. 2007. Report on fiber-reinforced polymer (FRP) reinforcement for concrete structures. ACI Committee 440, Farmington Hills, MI.

ACI 440.2R-08. 2008. Guide for the design and construction of externally bonded FRP systems for strengthening concrete structures. ACI Committee 440, Farmington Hills, MI.

Alkhrdaji, T., A. Nanni, and R. Mayo. 2000. Upgrading Missouri transportation infrastructure: Solid RC decks strengthened with FRP systems. *Transportation Research Record*, no. 1740:157–69.

Arduini, M., A. Tommaso, and A. Nanni. 1997. Brittle failure in FRP plate and sheet bonded beams. *ACI Structural Journal* 94 (4): 363–70.

Barbero, E. J. 2011. *Introduction to composite materials design*. 2nd ed. Boca Raton, FL: CRC Press, Taylor & Francis Group.

Beer, F. P., and E. R. Johnston Jr. 1992. *Mechanics of materials*. 2nd ed. New York: McGraw-Hill.

Bencardino, F., G. Spadea, and R. N. Swamy. 2002. Strength and ductility of reinforced concrete beams externally reinforced with carbon fiber fabric. *ACI Structural Journal* 99 (2): 163–71.

Decker, B. R. 2007. A method of strengthening monitored deficient bridges. MS thesis. Kansas State Univ., Manhattan, Kansas.

Ehsani, M. R. 1993. Glass fiber reinforcing bars. In *Alternative materials for the reinforcement and prestressing of concrete*, ed. J. L. Clarke. London: Blackie Academic & Professional.

Gibson, R. F. 1994. *Principles of composite material mechanics*. New York: McGraw Hill.

Gordon, J. E. 1976. *The new science of strong materials*. 2nd ed. Princeton, NJ: Princeton University Press.

Griffith, A. A. 1920. The phenomena of rupture and flow in solids. *Philosophical Transactions of the Royal Society* 221A:163–98.

Hawileh, R. A., M. Naser, W. Zaidan, and H. A. Rasheed. 2009. Modeling of insulated CFRP-strengthened reinforced concrete T-beam exposed to fire. *Engineering Structures* 31 (12): 3072–79.

Hyer, M. W. 1998. *Stress analysis of fiber reinforced composite materials*. New York: WCB/McGraw Hill.

Jones, R. M. 1975. *Mechanics of composite materials*. New York: Hemisphere Publishing.

Larson, K. H., R. J. Peterman, and H. A. Rasheed. 2005. Strength–fatigue behavior of FRP strengthened prestressed concrete T Beams. *ASCE Journal of Composites for Construction* 9 (4): 313–26.

Malvar, L. 1998. Durability of composites in reinforced concrete. In *Proceedings of the first international conference on durability of composites for construction*, ed. B. Benmokrane and H. Rahman, 361–72. Sherbrooke, QC: University of Sherbrooke, Dept. of Civil Engineering.

Mandell, J. F. 1982. Fatigue behavior of fiber-resin composites. In *Developments in reinforced plastics*. Vol. 2, 67-107. London: Applied Science Publishers.

Odagiri, T., K. Matsumoto, and H. Nakai. 1997. Fatigue and relaxation characteristics of continuous aramid fiber reinforced plastic rods. In *Third international symposium on non-metallic (FRP) reinforcement for concrete structures (FRPRCS-3)*. Vol. 2, 227–34. Tokyo: Japan Concrete Institute.

Rasheed, H. A. 1996. Behavior and strength of composite tubes considering delaminations and other defects. PhD diss. University of Texas at Austin.

Rasheed, H. A., R. R. Harrison, R. J. Peterman, and T. Alkhrdaji. 2010. Ductile strengthening using externally bonded and near surface mounted composite systems. *Composite Structures* 92 (10): 2379–90.

Rasheed, H. A., K. H. Larson, and R. J. Peterman. 2006. Analysis and design procedure for FRP-strengthened prestressed concrete T-girders considering strength and fatigue. *ASCE Journal of Composites for Construction* 10 (5): 419–32.

Reinhart, T. J. 1990. Polymer matrix composites. *Advanced Materials and Processes* 137 (1): 33.

Schwartz, M. M. 1984. *Composites materials handbook*. New York: McGraw-Hill.

Tamimi, A. K., R. Hawileh, J. Abdalla, and H. A. Rasheed. 2011. Effects of ratio of CFRP plate length to shear span and end anchorage on flexural behavior of SCC R/C beams. *ASCE Journal of Composites in Construction* 15 (5): 908–19.

Wu, W. 1990. Thermo-mechanical properties of fiber reinforced plastics (FRP) bars. PhD diss. West Virginia Univ., Morgantown, WV.

Wuertz, A. F. 2013. Strengthening rectangular beams with NSM steel bars and externally bonded GFRP. MS thesis, Kansas State Univ., Manhattan, KS.

Yamaguchi, T., Y. Kato, T. Nishimura, and T. Uomoto. 1997. Creep rupture of FRP rods made of aramid, carbon and glass fibers. In *Third international symposium on non-metallic (FRP) reinforcement for concrete structures (FRPRCS-3)*. Vol. 2, 179–86. Tokyo: Japan Concrete Institute.

Zweben, C. 1989. Static strength and elastic properties. In *Delaware composite design encyclopedia*. Part I: Mechanical behavior and properties of composite materials, ed. C. Zweben, H. T. Hahn, and T. W. Chou, 49–70. Lancaster, PA: Technomic Publishing.

4 Design Issues

4.1 OVERVIEW

The design equations and procedures presented in the next three chapters are based on the knowledge and principles put forth by ACI 318-11 (2011) and ACI 440.2R-08 (2008). Strain compatibility as well as force and moment equilibrium must be enforced. Strain at the concrete substrate, at the time of strengthening, also needs to be considered. FRP is not allowed to carry compressive stresses, according to ACI 440.2R-08. Therefore, FRP used in tension is ignored during moment-reversal calculations. Environmental factors need to be incorporated into the design, as shown in Chapter 3. Limiting strains in shear and column confinement are based on effective strains that are obtained experimentally. Strengthening levels are limited by the capacity of the section to carry reasonable upgraded loads in case of FRP loss during fire. This applies to shear and column confinement as it applies to flexure.

4.2 DESIGN PHILOSOPHY OF ACI 440.2R-08

A limit-state design philosophy is adopted by ACI 440.2R-08. This process defines load and resistance factors to various limit states considered, including

1. Ultimate limit state in flexure, shear, and fatigue.
2. Serviceability limit state in short- and long-term deflections and cracking.

For each limit state, different damage or failure modes and related design parameters are examined. It is worth mentioning that the load factors and load combinations follow the requirements of ACI 318-11 except for strengthening limits in case of fire. On the other hand, reduction factors for FRP were calibrated by ACI 440.2R-08 to produce a reliability index of 3.5 and above. However, the reliability index may be reduced to 3–3.5 in the case of high FRP ratio used with low steel ratio. This case is almost nonexistent due to the strengthening limits during fire. The values of the reliability index of FRP strengthened members were decided based on the work of Szerszen and Nowak (2003), which is mainly applicable to unstrengthened reinforced-concrete (RC) structures.

4.3 STRENGTHENING LIMITS DUE TO LOSS OF COMPOSITE ACTION

These limits are assigned to protect the member from failure if the FRP system is lost because of fire damage or vandalism. The bare section of member without strengthening should be able to resist certain levels of the new or upgraded loads without capacity deficiency, as described by Equation (9.1) of ACI 440.2R-08.

$$(\phi R_n)_{\text{existing}} \geq (1.1S_{DL} + 0.75S_{LL})_{\text{new}} \tag{4.1}$$

The dead-load factor is close to unity, since the estimate of the new or upgraded dead load can be determined fairly accurately. The live-load factor is selected to exceed the statistical mean of annual maximum live-load factor, given by ASCE-SEI 7-10 (2010) to be 0.5. Once the member survives the damage incident, the FRP is supposed to be repaired or reinstalled.

However, in cases where the live load is sustained on the member for an extended period of time, like the cases of library stacks and heavy storage warehouses where live load exceeds 150 lb/ft^2 (7.2 kN/m^2), the live load factor of 0.75 is increased to 1.0 in Equation (4.1).

4.4 FIRE ENDURANCE

The strengthening level of externally bonded FRP is typically limited by the fire endurance of the member with FRP. This is because FRP undergoes severe degradation in bond and mechanical properties at or beyond its glass transition temperature (T_g), which is typically equal to 140°F–180°F (60°C–82°C) (ACI 440.2R-08).

Deuring (1994) tested CFRP plated beams with and without insulation. He showed that the unprotected FRP-strengthened beams achieved around 81 minutes of fire endurance. Identical beams with CFRP protected by a 40-mm thick (1.57-in. thick) calcium-silicate insulation plates endured fire for a longer 146 minutes. It is important to note that bond between concrete and CFRP was lost within the first few minutes of fire exposure when CFRP was not protected.

Blontrock, Taerwe, and Vandevelde (2001) tested CFRP-strengthened beams protected with different insulation boards under full service load plus ISO 834 fire-curve exposure (1975). They observed that the best fire endurance can be accomplished if U-shaped insulation boards are installed to the soffit and sides of beams.

Williams et al. (2008) tested CFRP-strengthened RC T-beams with U-shaped vermiculite/gypsum (VG) insulation 25–38 mm thick (1–1.5 in. thick) subjected to uniformly distributed service load and ASTM E119 standard fire curve (2002). The strengthened RC T-beams were able to withstand the fire exposure for up to 4 hours. The results of this test and the three-dimensional (3-D) FE analysis of the same test (Hawileh et al. 2009) showed that the insulation controlled the temperature transferred to steel and FRP below the critical temperature values.

ACI 440.2R-08 allows the extension of the concepts established by ACI 216R-89 suggesting limits to maintain safety against collapse due to fire.

Steel and concrete strengths are reduced at elevated temperature according to ACI 216R-89, while FRP strength is ignored. The resistance of the member at elevated temperature $R_{n\theta}$ may be determined based on testing or ACI 216R-89 guidelines. Equation (9.2) of ACI 440.2R-08 needs to be satisfied:

$$R_{n\theta} = S_{DL} + S_{LL} \tag{4.2}$$

where S_{DL} and S_{LL} are the load effects of the upgraded loading due to the addition of FRP.

Glass transition temperature T_g is conservatively taken as the critical temperature below which FRP needs to be kept to sustain its functionality. More research is needed to accurately identify critical temperatures for different types of FRP, as seen in Figure 4.1 (Naser et al. 2014).

4.5 OVERALL STRENGTH OF STRUCTURES

It is the responsibility of the designer to make sure that the overall strength of structure is adequate under different upgraded loads for various strengthened members. Slab-punching shear, column capacity, and footing-bearing capacity must be satisfied when slabs and beams are upgraded. In most of the cases, this works out well due to the reduction in dead- and live-load factors required by ACI 318-11 and ACI 440.2R-08 compared to those required by early versions of ACI 318 code. An example on strengthening of a reinforced-concrete slab designed according to ACI 318-83 is presented in Chapter 5.

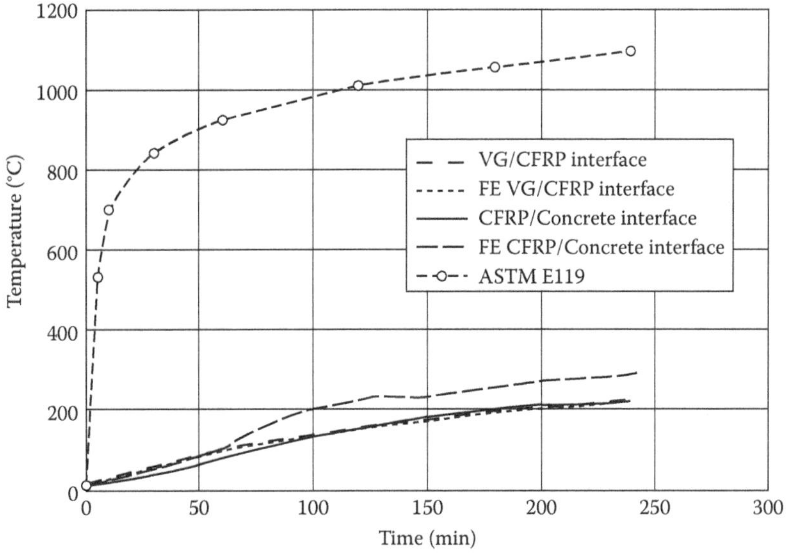

FIGURE 4.1 Experimental and numerical comparison of heat progression in CFRP-strengthened T-beams under fire (Naser et al. 2014. Courtesy of Multi-Science Publishing).

4.6 LOADING, ENVIRONMENTAL, AND DURABILITY FACTORS IN SELECTING FRP

4.6.1 Creep-Rupture and Fatigue

CFRP systems are highly tolerant to fatigue under cyclic loading and to creep rupture under sustained loading. GFRP systems are more vulnerable to these loading conditions. Accordingly, research has yielded that glass can sustain 0.3 times its ultimate strength, while aramid can sustain 0.5 of its ultimate strength and carbon can sustain 0.9 of its ultimate strength without a creep-rupture problem (Yamaguchi et al. 1997; Malvar 1998).

The stress level in FRP is computed under a total moment consisting of all sustained loading plus the maximum fatigue loading, as shown in Figure 4.2, using a linear elastic analysis. Values of sustained and cyclic stress levels are given in equation (4.3) by using a 0.6 reduction factor for the actual creep-rupture limits mentioned previously.

$$f_{f,s} = \begin{Bmatrix} 0.2 f_{fu} & \text{(for GFRP)} \\ 0.3 f_{fu} & \text{(for AFRP)} \\ 0.55 f_{fu} & \text{(for CFRP)} \end{Bmatrix} \tag{4.3}$$

4.6.2 Impact Resistance

GFRP and AFRP systems have higher impact resistance than CFRP systems. This may be attributed to the higher strain to failure they can sustain.

4.6.3 Acidity and Alkalinity

Dry carbon fiber is known to be resistant to alkalinity and acidity, while dry glass fiber is susceptible to degradation when subjected to these environments. On the

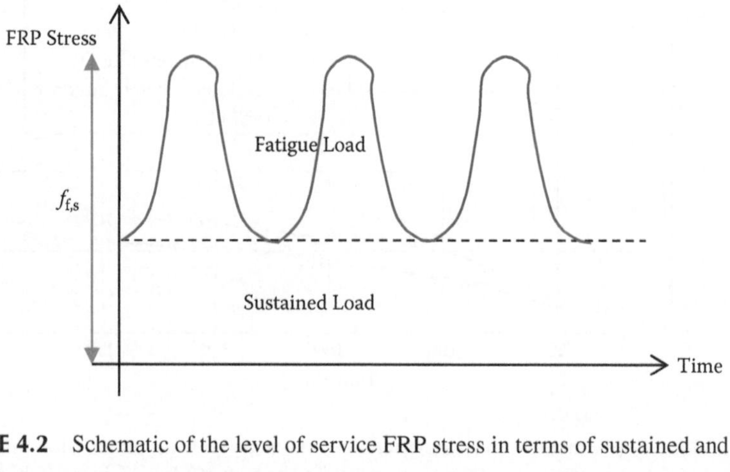

FIGURE 4.2 Schematic of the level of service FRP stress in terms of sustained and fatigue loading.

other hand, the resin matrix protects the fibers in general from alkalinity and acidity. However, it may only retard deterioration of glass fiber. Accordingly, carbon is the fiber of choice in applications having high alkalinity and high relative humidity.

4.6.4 Thermal Expansion

As stated previously, the coefficient of thermal expansion varies drastically between the fibers, matrix, and FRP system at various orientations. Along the fiber direction, CFRP has a coefficient of thermal expansion of zero while GFRP has a coefficient of thermal expansion comparable to concrete. The coefficient of thermal expansion of polymers is around five to seven times that of concrete. Thus, the transverse direction of an FRP system has much higher coefficients of thermal expansion. Differences in thermal expansion may potentially affect bond with concrete. However, this effect is limited for a temperature range of ±50°F (±28°C) (ACI 440.2R-08).

4.6.5 Electric Conductivity

GFRP and AFRP are known to be electric insulators. On the other hand, CFRP is conductive and should therefore be kept away from steel to avoid potential corrosion.

4.6.6 Durability

This is a subject of ongoing research. It is worth mentioning here that the use of FRP systems to completely seal the concrete surface is cautioned wherever moisture vapor transport is anticipated. Whenever possible, exposed concrete surfaces need to be allowed to enable moisture escape.

Chapter Problems

Problem 4.1

A library building has a simply supported rectangular concrete beam reinforced with four No. 6 bars in tension and two No. 3 bars in compression and No. 3 stirrups at 6 in. (152 mm) on center in shear. The details of the beam are shown in the following table. As part of the library upgrade, the beam is subjected to a 40% increase in live load, as shown in the table. Determine the adequacy of the bare-beam section to resist the upgraded loads in flexure and shear in case of fire.

Span	20 ft	6.1 m
B	12 in.	305 mm
H	20 in.	508 mm
f'_c	4 ksi	27.6 MPa
f_y	60 ksi	414 MPa
Top bars	#3	ϕ = 9.5 mm
Main bars	#6	ϕ = 19 mm
W_{DL}	0.9 k/ft	13.14 kN/m
W_{LL}	0.9 k/ft	13.14 kN/m

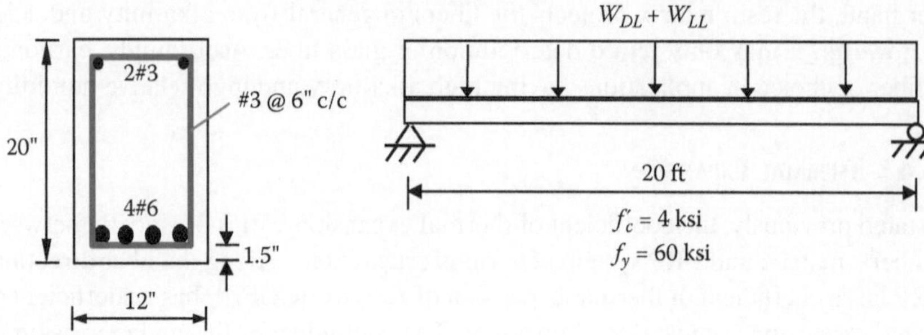

FIGURE 4.P.1

Problem 4.2

Derive the following expression of the elastic depth to the neutral axis of the cracked section using linear elastic cracked section analysis considering the contribution of sectional FRP:

$$k = \sqrt{\left(\frac{E_s}{E_c}\rho_s + \frac{E_f}{E_c}\rho_f\right)^2 + 2\left(\frac{E_s}{E_c}\rho_s + \frac{E_f}{E_c}\rho_f\frac{d_f}{d}\right)} - \left(\frac{E_s}{E_c}\rho_s + \frac{E_f}{E_c}\rho_f\right)$$

Problem 4.3

Derive the following expression for the elastic stress level in steel under service load considering the contribution of sectional FRP. Note that this expression is given by Equation (10.14) of ACI 440.2R-08:

$$f_{s,s} = \frac{\left[M_s + \varepsilon_{bi} A_f E_f \left(d_f - \frac{kd}{3}\right)\right](d - kd)E_s}{A_s E_s \left(d - \frac{kd}{3}\right)(d - kd) + A_f E_f \left(d_f - \frac{kd}{3}\right)(d_f - kd)}$$

Problem 4.4

Derive the following expression for the elastic stress level in FRP under service load considering the contribution of sectional steel and FRP. Note that this expression is given by Equation (10.15) of ACI 440.2R-08.

$$f_{f,s} = f_{s,s}\frac{E_f}{E_s}\frac{d_f - kd}{d - kd} - \varepsilon_{bi} E_f$$

Problem 4.5

Determine the elastic stress level in steel of the beam in Problem 4.1. Note that 50% of the original live load existed on the beam during strengthening. The beam was strengthened using two plies of CFRP with the properties listed in the following table. Compare the steel stress level to the limit set by Equation (10.6) of ACI 440.2R-08.

Material	Modulus of Elasticity E, ksi (GPa)	Ultimate Strain ε_u, in./in. (mm/mm)	Ultimate Strength f_u, ksi (MPa)	Ply Thickness t_f, in. (mm)
Carbon FRP (sheets)	33,000 (227)	0.014 (0.014)	462 (3187)	0.0065 (0.165)

Problem 4.6:

For the beam in Problems 4.1 and 4.5, determine the elastic stress level in FRP. Note that 50% of the original live load existed on the beam during strengthening. The beam was strengthened using two plies of CFRP with the properties listed in problem 4.5. Compare the FRP stress level to the limit set by Equation (4.3) of this chapter.

REFERENCES

ACI216R-89. 1989. Guide for determining the fire endurance of concrete elements (reapproved 2001). ACI Committee 216, Farmington Hills, MI.

ACI 318-11. 2011. Building code requirements for structural concrete and commentary. ACI Committee 318, Farmington Hills, MI.

ACI 440.2R-08. 2008. Guide for the design and construction of externally bonded FRP systems for strengthening concrete structures. ACI Committee 440, Farmington Hills, MI.

ASCE-SEI 7-10. 2010. Minimum design loads for buildings and other structures. American Society of Civil Engineers and the Structural Engineering Institute, Reston, VA.

ASTM E119. 2002. Standard test methods for fire tests of building construction and materials. ASTM International, West Conshohocken, PA.

Blontrock, H., L. Taerwe, and V. Vandevelde. 2001. Fire testing of concrete slabs strengthened with fibre composite laminates. In *Proceedings of the fifth annual symposium on fibre-reinforced-plastic reinforcement for concrete structures (FRPRCS-5)*, ed. C. Burgoyne, 547–56. London: Thomas Telford.

Deuring, M. 1994. Brandversuche an Nachtraglich Verstarkten Tragern aus Beton. Research Report EMPA No. 148,795. Swiss Federal Laboratories for Materials Testing and Research, Dubendorf, Switzerland.

ISO-834. 1975. Fire resistance tests: Elements of building construction. International Standards Organisation, Geneva.

Malvar, L. 1998. Durability of composites in reinforced concrete. In *Proceedings of the first international conference on durability of composites for construction*, ed. B. Benmokrane and H. Rahman, 361–72. Sherbrooke, QC: University of Sherbrooke, Dept. of Civil Engineering.

Naser, M. Z., Hawileh, R. A., and Rasheed, H. A., "Performance of RC T-Beams Externally Strengthened with CFRP Laminates under Elevated Temperatures," *Journal of Structural Fire Engineering*, Multi-Science Publishing, Vol. 5, No. 1, March 2014, pp. 1–24.

Szerszen, M. M., and A. S. Nowak. 2003. Calibration of design code for buildings (ACI 318): Part 2—Reliability analysis and resistance factors. *ACI Structural Journal* 100 (3): 383–91.

Williams, B., V. K. R. Kodur, M. F. Green, and L. Bisby. 2008. Fire endurance of fiber-reinforced polymer strengthened concrete T-beams. *ACI Structural Journal* 105 (1): 60–67.

Yamaguchi, T., Y. Kato, T. Nishimura, and T. Uomoto. 1997. Creep rupture of FRP rods made of aramid, carbon and glass fibers. In *Proceedings of the third international symposium on non-metallic (FRP) reinforcement for concrete structures (FRPRCS-3)*. Vol. 2, 179–86. Tokyo: Japan Concrete Institute.

5 Flexural Strengthening of Beams and Slabs

5.1 OVERVIEW

External bonding of carbon FRP reinforcement to the tension face of a concrete beam to increase its flexural capacity has been introduced by Meier (1987). Since then, a large volume of literature has been added to qualify this strengthening technique. The extension of this technique to near-surface-mounted FRP bars and strips should also be noted (Alkhrdaji, Nanni, and Mayo 2000). These strengthening techniques resulted in flexural capacity increases of 20%–200% compared to the unstrengthened beams. However, the strengthening limits imposed by ACI 440.2R-08 (2008) restrict the strengthening ratio up to about 40%–50%. More research on the insulation of FRP under fire is needed to waive the need for these strengthening limitations and allow higher strengthening ratios to be implemented.

5.2 STRENGTH REQUIREMENTS

It is widely established that the design nominal flexural strength of a strengthened section reduced by the strength reduction factor should be equal to or greater than the factored moment of the upgraded loads:

$$\phi \bar{M}_n \geq \bar{M}_u \tag{5.1}$$

The nominal strengthened moment capacity $(\bar{M}_n)$ with steel reinforcement and FRP is to be determined using

- Strain compatibility
- Internal force equilibrium
- Internal moment equilibrium
- Expected failure mode

5.3 STRENGTH REDUCTION FACTORS

The strength reduction factor of a strengthened section is given by Equation (10.5) of ACI 440.2R-08 based on the requirements of ACI 318-05 (2005). This factor ϕ is a function of the net tensile strain in the most extreme tension steel layer at nominal strength:

$$\phi = \begin{cases} 0.9 & \text{for} \quad \varepsilon_t \geq 0.005 \\ 0.65 + \dfrac{0.25(\varepsilon_t - \varepsilon_{sy})}{0.005 - \varepsilon_{sy}} & \text{for} \quad \varepsilon_{sy} < \varepsilon_t < 0.005 \\ 0.65 & \text{for} \quad \varepsilon_t \leq \varepsilon_{sy} \end{cases} \tag{5.2}$$

The above values are set to $\phi = 0.9$ for a tension-controlled failure (ductile) and to $\phi = 0.65$ for a compression-controlled failure (brittle without yielding), while ϕ is determined from the linear interpolation between the two extremes in the transition zone, as seen in Figure 5.1.

For Grade 60 reinforcement, $\varepsilon_{sy} = \frac{60\text{ ksi}}{29{,}000\text{ ksi}} \approx 0.002$. Thus for $\phi = 0.9$, $\frac{c}{d_t} \leq 0.375$, and for $\phi = 0.65, \frac{c}{d_t} \geq 0.6$. In the transition, $\phi = 0.65 + 0.25\left[\frac{d_t}{c} - \frac{5}{3}\right]$. In addition to the factor ϕ, another strength reduction factor for FRP (Ψ_f) is multiplied by the flexural contribution of FRP reinforcement ($\bar{M}_{nf}$), as shown in Equation (10.13) of the ACI 440.2R-08:

$$\bar{M}_n = A_s f_s \left(d - \frac{\beta_1 c}{2} \right) + \Psi_f A_f f_{fe} \left(d_f - \frac{\beta_1 c}{2} \right) \tag{5.3}$$

This Ψ_f factor is taken as 0.85 based on the reliability analysis of the experimentally calibrated statistical values to mainly account for the less predictable failure mode of delamination of FRP reinforcement (Okeil, Bingol, and Alkhrdaji 2007).

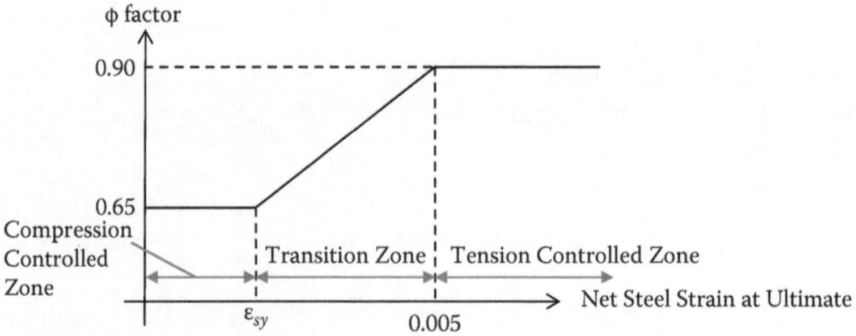

FIGURE 5.1 Variation of ϕ factor as a function of the extreme net steel strain.

5.4 FLEXURAL FAILURE MODES

For FRP-strengthened reinforced concrete section, the following failure modes are likely to take place in flexure (GangaRao and Vijay 1998):

1. Ductile crushing of concrete: crushing of concrete in compression after yielding of tensile steel (desirable failure mode)
2. Brittle crushing of concrete: crushing of concrete in compression before yielding of tensile steel (undesirable failure mode)
3. Rupture of FRP: rupture of FRP reinforcement after yielding of tensile steel (desirable failure mode)
4. Cover delamination: Shear/tension delamination of the concrete cover starting at the FRP curtailment (not covered by ACI 440.2R-08)
5. Debonding of FRP: The intermediate crack-induced debonding of FRP reinforcement (covered by ACI 440.2R-08)

5.4.1 Ductile Crushing of Concrete

This is the failure mode in which concrete reaches the limit of useful compressive strain of 0.003 after yielding of tension reinforcement $\varepsilon_s > \varepsilon_{sy}$. It is a desirable failure mode because it involves a warning sign represented by the ductile yielding of steel. It competes with the debonding failure mode, which alternatively takes place in sections moderately strengthened with FRP. If the FRP is anchored with transverse FRP U-wraps or stirrups, this failure mode would be the dominant one, since debonding and cover delamination are prevented or delayed by the FRP U-wraps (Rasheed et al. 2010).

5.4.1.1 Flexural Strengthening of a Singly Reinforced Section

For the concrete crushing failure mode after yielding of primary steel reinforcement (Figure 5.2):

$$\text{Strain compatibility: } \frac{\varepsilon_{cu}}{c} = \frac{\varepsilon_{cu} + \bar{\varepsilon}_f}{d_f}, \text{ where } \bar{\varepsilon}_f = \varepsilon_{fe} + \varepsilon_{bi} \quad (5.4)$$

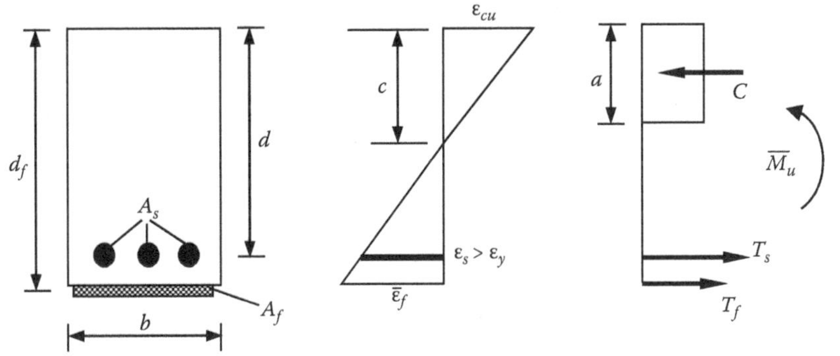

FIGURE 5.2 Singly reinforced cross section with strain distribution and force profile.

$$\text{Force equilibrium: } C = T_s + T_f \Rightarrow 0.85 f_c' b \beta_1 c = A_s f_y + A_f E_f \varepsilon_{fe} \tag{5.5}$$

$$\text{Moment equilibrium: } \bar{M}_u = \bar{\phi} T_s jd_s + \bar{\phi} T_f jd_f = \bar{\phi} A_s f_y \left(d - \frac{\beta_1 c}{2} \right)$$

$$+ \bar{\phi} \Psi_f A_f E_f \varepsilon_{fe} \left(d_f - \frac{\beta_1 c}{2} \right) \tag{5.6}$$

where ε_{bi} is the strain in the extreme fiber of concrete substrate at the time of strengthening.

Design solution approach by ACI 440.2R-08:

1. Assume A_f
2. Assume c and iterate to satisfy force equilibrium and strain compatibility
3. Compute $\phi \bar{M}_n$ and compare it with $\bar{M}_{u\text{-}required}$
4. Adjust A_f and repeat steps 1–3 until $\phi \bar{M}_n \geq \bar{M}_{u\text{-}required}$

Design solution approach by direct equations:
The design solution for this failure mode is possible to determine in closed form (Rasheed and Pervaiz 2003):

$$\frac{\varepsilon_{cu}}{c} = \frac{\varepsilon_{cu} + \bar{\varepsilon}_f}{d_f} \Rightarrow c = \frac{0.003 d_f}{0.003 + \bar{\varepsilon}_f} = \frac{a}{\beta_1} \tag{5.7}$$

$$\varepsilon_{fe} = \bar{\varepsilon}_f - \varepsilon_{bi} \leq \varepsilon_{fd} \tag{5.8}$$

where ε_{fd} is the debonding strain limit of FRP

$$0.85 f_c' ba = A_s f_y + A_f f_{fe} \tag{5.9}$$

Solving for A_f in Equation (5.9):

$$A_f = \frac{0.85 f_c' ba - A_s f_y}{f_{fe}} \tag{5.10}$$

$$\rho_f = 0.85 \frac{f_c'}{f_{fe}} \left(\frac{a}{d} \right) - \rho_s \frac{f_y}{f_{fe}} \tag{5.11}$$

where $\rho_f = \dfrac{A_f}{bd}$, $\rho_s = \dfrac{A_s}{bd}$

$$\bar{M}_u = \bar{\phi} \bar{M}_n = \bar{\phi} A_s f_y \left(d - \frac{a}{2} \right) + \bar{\phi} \Psi_f A_f f_{fe} \left(d_f - \frac{a}{2} \right) \tag{5.12}$$

Dividing Equation (5.12) by $\bar{\phi} f_c' b d_f^2$,

$$\frac{\bar{M}_u}{\bar{\phi} f_c' b d_f^2} = \rho_s \frac{f_y}{f_c'} \frac{d}{d_f^2}\left(d - \frac{a}{2}\right) + \Psi_f \rho_f \frac{f_{fe}}{f_c'} \frac{d}{d_f^2}\left(d_f - \frac{a}{2}\right) \tag{5.13}$$

Substituting Equation (5.11) into (5.13) and rearranging the terms,

$$\left(\frac{a}{d_f}\right)^2 - \left[2 - 1.176 \frac{1-\Psi_f}{\Psi_f} Q_1\right]\frac{a}{d_f} + \frac{2.353}{\Psi_f} Q_2 = 0 \tag{5.14}$$

where

$$Q_1 = \rho_s \frac{f_y}{f_c'} \frac{d}{d_f} \tag{5.15}$$

$$Q_2 = \frac{\bar{M}_u}{\bar{\phi} f_c' b d_f^2} + Q_1\left(\Psi_f - \frac{d}{d_f}\right) \tag{5.16}$$

When $\Psi_f = 0.85$ is substituted into Equation (5.14),

$$\left(\frac{a}{d_f}\right)^2 - [2 - 0.208 Q_1]\frac{a}{d_f} + 2.77 Q_2 = 0 \tag{5.17}$$

or

$$\frac{a}{d_f} = \frac{1}{2}\left[(2 - 0.208 Q_1) - \sqrt{(2 - 0.208 Q_1)^2 - 11.08 Q_2}\right] \tag{5.18}$$

Now, to establish a range of values for Q_1,

$f_c' = 3000 - 8000$ psi

$f_y = 40 - 75$ ksi

$$\rho_{s,\min} = \frac{3\sqrt{3{,}000}}{75{,}000} = 2.19 \times 10^{-3} < \frac{200}{75{,}000} = 2.67 \times 10^{-3} \quad \text{controls}$$

or

$$\rho_{s,\min} = \frac{3\sqrt{8{,}000}}{40{,}000} = 6.71 \times 10^{-3} > \frac{200}{40{,}000} = 5.0 \times 10^{-3}, \quad 6.71 \times 10^{-3} \text{controls}$$

$$\rho_{s,\max} = 0.75\rho_b = 0.75 \times 0.85\beta_1 \frac{f_c'}{f_y} \frac{87}{87 + f_y} \quad \textit{for } \varepsilon_{st} \approx 0.004 \text{ (ductile failure)}$$

$$= 0.75 \times 0.85 \times 0.65 \times \frac{8{,}000}{40{,}000} \frac{87}{87 + 40} = 0.0568$$

or

$$= 0.75 \times 0.85 \times 0.85 \times \frac{3{,}000}{75{,}000} \frac{87}{87+75} = 0.01164$$

$$\frac{d}{d_f} = 0.8 - 0.9$$

$$Q_1 = 0.01164 \times \frac{75}{3} \times 0.9 = 0.262 \text{ controls (upper bound)}$$

$$Q_1 = 2.67 \times 10^{-3} \times \frac{75}{3} \times 0.8 = 0.0534$$

$$Q_1 = 0.0568 \times \frac{40}{8} \times 0.9 = 0.2556$$

$$Q_1 = 5.0 \times 10^{-3} \times \frac{40}{8} \times 0.8 = 0.0268 \text{ controls (lower bound)}$$

$$0.0268 < Q_1 < 0.262 \Rightarrow B = 1.946 - 1.994 \text{ close to 2}$$

where $B = 2 - 0.208Q_1$

Assuming $B = 2 \Rightarrow$

$$\frac{a}{d_f} = 1 - \sqrt{1 - 2.77Q_2} \Rightarrow \text{ for } \Psi_f = 0.85 \tag{5.19}$$

$$\frac{a}{d_f} = 1 - \sqrt{1 - 2.35Q_2} \Rightarrow \text{ for } \Psi_f = 1.0 \tag{5.20}$$

Example 5.1: Design

Solve Example 15.3 of ACI 440.2R-08 using the direct approach (Figure 5.3). The problem statement from the design guide goes as follows: A simply supported reinforced concrete beam is located in an unoccupied warehouse. It is required to be strengthened by a 50% increase in its live-load carrying capacity.

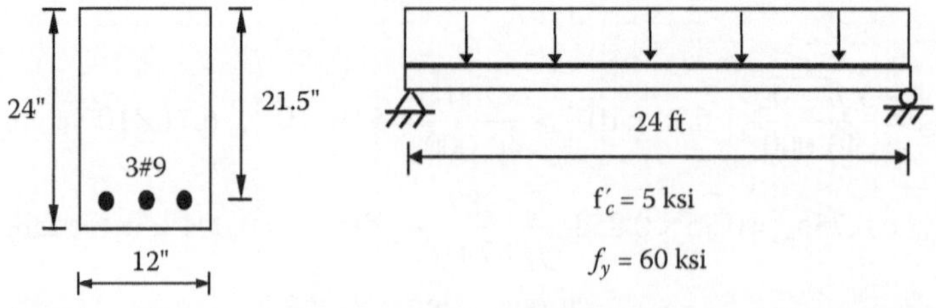

FIGURE 5.3 Example 15.3 of ACI 440.2R-08 showing the beam profile and the cross-section details.

TABLE 5.1
Beam Parameters of Example 15.3 of ACI 440.2R-08

Span	24 ft	7.32 m
B	12 in.	305 mm
H	24 in.	609.6 mm
d	21.5 in.	546 mm
f'_c	5 ksi	34.5 MPa
f_y	60 ksi	414 MPa
Main bars	#9	$\phi = 28.6$ mm
ϕM_n	266 k-ft	361 kN-m

The beam is assumed to have sufficient shear strength and adequate deflection and crack-control serviceability requirements while resisting the additional loads. Design a flexural strengthening system to resist the upgraded moment $\bar{M}_u = 294.4\,k-ft = 399\,kN-m$. The beam parameters are given in Table 5.1.

Solution:

Direct approach:

Use $d_f = h = 24$ in.

$$\bar{M}_u = 294.4\,k-ft$$

$$\rho_s = \frac{A_s}{bd} = \frac{3in^2}{12\times 21.5} = 0.01163$$

$$Q_1 = \rho_s \frac{f_y}{f'_c}\frac{d}{d_f} = 0.01163\times\frac{60}{5}\times\frac{21.5}{24} = 0.125$$

$$Q_2 = \frac{\bar{M}_u}{\phi f'_c b d_f^2} + Q_1\left(\Psi_f - \frac{d}{d_f}\right) = \frac{294.4\times 12}{0.9\times 5\times 12\times 24^2} + 0.125\times\left(0.85 - \frac{21.5}{24}\right) = 0.1079$$

$$\left(\frac{a}{d_f}\right)^2 - \left[2 - 1.176\times\frac{0.15}{0.85}\times 0.125\right]\frac{a}{d_f} + \frac{2.353}{0.85}\times 0.1079 = 0$$

$$\frac{a}{d_f} = 1 - \sqrt{1 - 2.77\times 0.1079} = 0.163$$

$$a = 3.9 \text{ in}$$

$$\bar{\varepsilon}_f = \frac{0.003 d_f}{\frac{a}{\beta_1}} - 0.003 = \frac{0.003\times 24}{\frac{3.9}{0.8}} - 0.003 = 0.01177$$

$$\varepsilon_{fe} = \bar{\varepsilon}_f - \varepsilon_{bi} = 0.01177 - 0.00061 = 0.01116$$

$$f_{fe} = E_f \varepsilon_{fe} = 5360 \times 0.01116 = 59.82 \text{ ksi}$$

$$\rho_f = 0.85 \frac{f_c'}{f_{fe}} \left(\frac{a}{d} \right) - \rho_s \frac{f_y}{f_{fe}} = 0.85 \times \frac{5}{59.82} \times \frac{3.9}{21.5} - 0.01163 \times \frac{60}{59.82} = 0.001223$$

$$A_f = 0.001223 \times 12 \times 21.5 = 0.316 \text{ in}^2$$

$$b_f = \frac{A_f}{t_f} = \frac{0.316}{0.04} = 7.9 \text{ in. } \Rightarrow \text{Use 8" with one layer of FRP}$$

$$\varepsilon_{fd} = 0.083 \sqrt{\frac{5000 \text{ psi}}{1(5{,}360{,}000 \text{ psi})(0.04)}} = 0.0127 > \varepsilon_{fe} = 0.01116$$

$\Rightarrow$ No Debonding $\Rightarrow$ O.K.

ACI 440.2R-08 used 2 layers of FRP of full beam width ($A_f = 0.96$ in^2)

Example 5.2: Analysis

Use the results of Example 5.1 to check the moment capacity of the strengthened section $\bar{M}_u = 294.4 k - ft = 399 kN - m$.

Solution:

Using force equilibrium,

$$0.85 f_c' b \beta_1 c = A_s f_y + A_f E_f \varepsilon_{fe} = A_s f_y + A_f E_f \left(\frac{0.003 d_f}{c} - 0.003 - \varepsilon_{bi} \right)$$

$$40.8c = 3 \times 60 + 0.32 \times 5360 \times \left(\frac{0.072}{c} - 0.00361 \right)$$

$$40.8c^2 - 173.808c - 123.49 = 0$$

$$c = \frac{173.808 \pm \sqrt{173.808^2 + 4 \times 40.8 \times 123.49}}{2 \times 40.8} = 4.88 \, in \text{ (positive root)}$$

$$a = 3.904, \; \varepsilon_{fe} = 0.011144 \Rightarrow f_f = 59.73 \text{ ksi}$$

$$\bar{M}_u = 0.9 \times \left[3 \times 60 \left(21.5 - \frac{3.904}{2} \right) + 0.85 \times 0.32 \times 59.73 \times \left(24 - \frac{3.904}{2} \right) \right]$$

$$= 3489.16 \text{ k-in} = 290.76 \text{ k-ft} \approx 294.4 \text{ k-ft}$$

Why the small difference? Because of rounding the constants of Equation(5.19)

Check the steel strain:

$$\varepsilon_s = \frac{0.003}{4.88}(21.5 - 4.88) = 0.0102 > 0.005 \Rightarrow \phi = 0.9 \Rightarrow O.K.$$

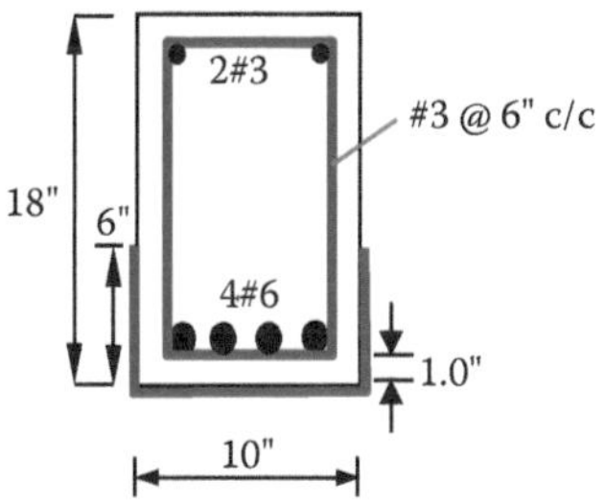

FIGURE 5.4 Example 5.3 showing the cross-section details.

Example 5.3: Analysis

For the beam shown in Figure 5.4, determine the increase in moment-carrying capacity $\frac{\bar{M}_n}{M_n}$. Neglect the compression reinforcement. The CFRP covers the tension face of the beam and is wrapped 6" up the two sides of the beam section, as shown (Rasheed et al. 2010). The CFRP sheets have the following properties:

M-Brace CF130
Two plies

$f_{fu} = 440$ ksi based on fiber net area

$\varepsilon_{fu} = 0.014$

$E_f = 33{,}000$ ksi

$t_{sheet} = 0.0065$ in

B	10 in.	254 mm
H	18 in.	457.2 mm
f'_c	5 ksi	34.5 MPa
f_y	83.5 ksi	576 MPa
Main bars	#6	$\phi = 19$ mm
Cover	1 in.	25.4 mm

Solution:

Since these examples are intended to target ductile crushing failure, there is no need to use the environmental factors (C_e) of ACI 440.2R-08.

$$d = d_t = 18 - 1'' - \frac{3}{8} - \frac{1}{2} \times \frac{6}{8} = 16.25''$$

$$A_s = 4 \times 0.44 = 1.76 \text{ in}^2$$

$$d_f = \frac{2 \times (10 \times t \times 18 + 2 \times 6 \times t \times 15)}{2 \times (10t + 2 \times 6t)} = 16.36''$$

$$A_f = (10 \times 0.0065'' + 2 \times 6 \times 0.0065) \times 2 \text{ plies} = 0.286 \text{ in}^2$$

Assume ductile crushing failure mode:

$$c = \frac{A_s f_y + A_f f_{fe}}{0.85 f'_c \beta_1 b} \qquad \text{Equation (10.12) of ACI 440.2R-08}$$

Using Equation (2.17):

$$\beta_1 = 1.05 - 0.05 \times \frac{5000}{1000} = 0.8$$

$\varepsilon_{bi} = 0$ The beam was not loaded during strengthening (Rasheed et al. 2010)

$$\varepsilon_{fe} = 0.003 \frac{d_f - c}{c} - \varepsilon_{bi} \qquad \text{Equation (10.3) of ACI 440.2R-08}$$

$$= 0.003 \frac{16.36 - c}{c}$$

$$f_{fe} = E_f \varepsilon_{fe} = 33{,}000 \times 0.003 \frac{16.36 - c}{c} = 99 \frac{16.36 - c}{c}$$

Force equilibrium:

$$0.85 \times 5 \times 0.8 \times 10c = 1.76 \times 83.5 + 0.286 \times 99 \left(\frac{16.39 - c}{c} \right)$$

Multiply the equation by c:

$$34c^2 - 146.96c - 463.22 + 28.31c = 0$$

$$34c^2 - 118.65c - 463.22 = 0$$

$$c = \frac{118.65 + \sqrt{118.65^2 - 4 \times 34 \times (-463.22)}}{2 \times 34} = 5.83 \text{ in}$$

$$\varepsilon_{fe} = 0.003 \frac{16.36 - 5.83}{5.83} = 0.00542 \le \varepsilon_{fd}$$

$$\varepsilon_{fd} = 0.083 \sqrt{\frac{f'_c}{nE_f t_f}} = 0.083 \sqrt{\frac{5000}{2 \times 33 \times 10^6 \times 0.0065}} = 0.00896 \le 0.9 C_e \varepsilon_{fu} = 0.01197$$

$$\varepsilon_s = 0.003 \frac{d - c}{c} = 0.003 \frac{16.25 - 5.83}{5.83} = 0.00536 > 0.005 \Rightarrow \bar{\phi} = 0.9$$

Ductile crushing failure confirmed

$$f_{fe} = 33{,}000 \times 0.00542 = 178.86 \text{ ksi}$$

$$\bar{M}_n = A_s f_y \left(d - \frac{\beta_1 c}{2} \right) + \Psi_f A_f f_{fe} \left(d_f - \frac{\beta_1 c}{2} \right)$$

$$\bar{M}_n = 1.76 \times 83.5 \times \left(16.25 - \frac{0.8 \times 5.83}{2} \right)$$

$$+ 0.85 \times 0.286 \times 178.86 \times \left(16.36 - \frac{0.8 \times 5.83}{2} \right)$$

$$\bar{M}_n = 2655.34 \text{k-in.} = 221.30 \text{ k-ft}$$

Compare to 247.6 k-ft from experiment.

Why the difference? Contribution of compression steel and steel strain hardening in tension as well as $\Psi_f = 0.85$.

Unstrengthened beam capacity (M_n):

$$c = \frac{1.76 \times 83.5}{0.85 \times 5 \times 0.8 \times 10} = 4.32 \text{ in.}$$

$$\frac{c}{d_t} = \frac{4.32}{16.25} = 0.266 < 0.375 \Rightarrow \text{ tension controlled failure} \Rightarrow \phi = 0.9$$

$$M_n = A_s f_y\left(d - \frac{\beta_1 c}{2}\right) = 1.76 \times 83.5 \times \left(16.25 - \frac{0.8 \times 4.32}{2}\right) = 2134.15 \text{ k-in.}$$

$$= 177.85 \text{ k-ft}$$

Compare to 180.4 k-ft from the control beam in experiment.

Why the small difference? Contribution of compression steel and steel strain hardening in tension.

$$\text{S.R.} = \frac{\bar{M}_n}{M_n} = \frac{221.3}{177.85} = 1.244$$

From the experiments:

$$\text{S.R.} = \frac{247.6}{180.4} = 1.373$$

Example 5.4: Design

For the beam given in Example 5.3, determine the CFRP area needed to increase the total section moment capacity by 25%.

Solution:

Unstrengthened beam capacity:

$$M_u = 0.9 \times 2134.15 \text{ k-in} = 1920.74 \text{ k-in}$$

$$\bar{M}_u = M_u \times 1.25 = 2400.92 \text{ k-in}$$

Assume $d_f = h = 18''$ (CFRP sheets bonded only to soffit).

$$\rho_s = \frac{A_s}{bd} = \frac{1.76}{10 \times 16.25} = 0.01083$$

$$Q_1 = \rho_s \frac{f_y}{f_c'} \frac{d}{d_f} = 0.01083 \times \frac{83.5}{5} \times \frac{16.25}{18} = 0.1633$$

$$Q_2 = \frac{\bar{M}_u}{\phi f_c' b d_f^2} + Q_1\left(\Psi_f - \frac{d}{d_f}\right) = \frac{2400.92}{0.9 \times 5 \times 10 \times 18^2} + 0.1633\left(0.85 - \frac{16.25}{18}\right) = 0.156$$

$$\left(\frac{a}{d_f}\right)^2 - (2-0.208Q_1)\frac{a}{d_f} + 2.77Q_2 = 0$$

$$\left(\frac{a}{d_f}\right)^2 - 1.966\frac{a}{d_f} + 0.4321 = 0$$

$$\frac{a}{d_f} = \frac{1.966 - \sqrt{1.966^2 - 4\times 0.4321}}{2} = 0.252$$

$a = 4.54$ in

$$\varepsilon_s = 0.003\frac{16.25 - \frac{4.54}{0.8}}{\frac{4.54}{0.8}} = 0.00559 > 0.005$$

$$\bar{\varepsilon}_f = \frac{0.003d_f}{\frac{a}{\beta_1}} - 0.003 = 0.00652$$

$\varepsilon_{fe} = \bar{\varepsilon}_f - \varepsilon_{bi} = 0.00652 < \varepsilon_{fd} = 0.00896$ (assuming 2 layers of FRP).

$f_{fe} = 33{,}000 \times 0.00652 = 215.15$ ksi

$$\rho_f = 0.85\frac{f_c'}{f_{fe}}\left(\frac{a}{d}\right) - \rho_s\frac{f_y}{f_{fe}} = 0.85 \times \frac{5}{215.15} \times \frac{4.54}{16.25} - 0.01083 \times \frac{83.5}{215.15} = 0.001313$$

$A_f = 0.001313 \times 10 \times 16.25 = 0.213$ in^2

Why is it smaller than 0.286 in^2 in Example 5.3?

Because d_f is larger (18").

$$b_f = \frac{A_f}{nt_f} = \frac{0.213}{2\times 0.0065} = 16.42 \text{ in} > b = 10'' \text{ N.G.}$$

Either wrap around the sides or use 3 layers.

Using four layers and a partial layer:

$3 \times 10 \times 0.0065 + b_f' \times 0.0065 = 0.213$

$b_f' = 2.77$ in. ≈ 3.0 in.

$$\text{check } \varepsilon_{fd} = 0.083\sqrt{\frac{5000}{3\times 33\times 10^6 \times 0.0065}} = 0.00732 > \varepsilon_{fe} \text{ O.K.}$$

The design in Figure 5.5 is more economical (less FRP area) than the wrapped design in Example 5.3. However, the wrapped design has its advantages in confining the cover area and reducing the interfacial shear stresses.

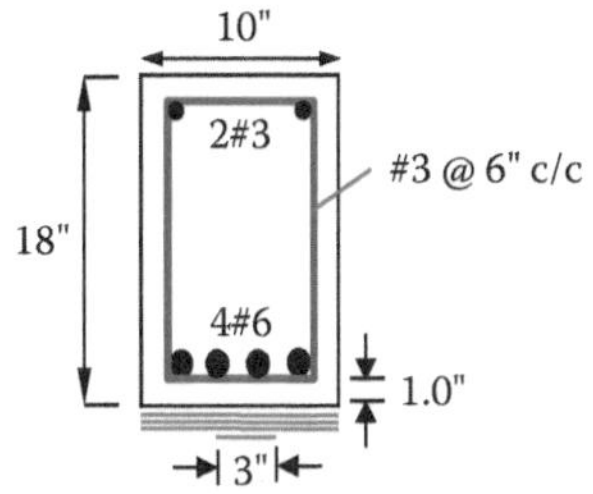

FIGURE 5.5 Example 5.4 showing the FRP design details.

Example 5.5: Design

Redo the strengthening design of Example 5.4 using near-surface mounted (NSM) CFRP tape (Figure 5.6) with the following properties:

Aslan 500
Dimensions: 0.63 in. x 0.079 in.
$f_{fu} = 300$ ksi
$E_f = 18{,}000$ ksi
$\varepsilon_{fu} = 0.0167$

$\bar{M}_u = 2400.92$ k-in

$$d_f = h - \frac{b_f}{2} = 18'' - \frac{0.63}{2} = 17.69''$$

$$Q_1 = \rho_s \frac{f_y}{f_c'} \frac{d}{d_f} = 0.01083 \times \frac{83.5}{5} \times \frac{16.25}{17.69} = 0.166$$

$$Q_2 = \frac{2400.92}{0.9 \times 5 \times 10 \times 17.69^2} + 0.166\left(0.85 - \frac{16.25}{17.69}\right) = 0.159$$

$$\left(\frac{a}{d_f}\right)^2 - (2 - 0.208 \times 0.166)\frac{a}{d_f} + 2.77 \times 0.159 = 0$$

$$\left(\frac{a}{d_f}\right)^2 - 1.965\frac{a}{d_f} + 0.4404 = 0$$

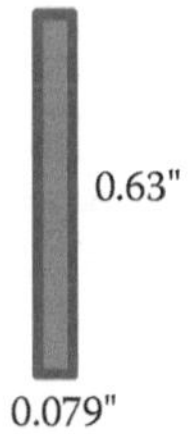

FIGURE 5.6 CFRP strip for NSM application.

$B \approx 2 \Rightarrow$

$$\frac{a}{d_f} = 1 - \sqrt{1 - 0.4404} = 0.252$$

$$a = 4.46 \text{ in} \Rightarrow c = 5.57 \text{ in}$$

$$\varepsilon_s = 0.003\frac{16.25 - 5.57}{5.57} = 0.0057 > 0.005 \Rightarrow \text{ ductile crushing}$$

$$\bar{\varepsilon}_f = \frac{0.003 \times 17.69}{5.57} - 0.003 = 0.00652$$

$$\varepsilon_{fe} = \bar{\varepsilon}_f - \varepsilon_{bi} = 0.00652 < \varepsilon_{fd} = 0.7\varepsilon_{fu} = 0.7 \times 0.95 \times 0.0167 = 0.0111$$

$$f_{fe} = 18{,}000 \times 0.00652 = 117.36 \text{ ksi}$$

$$\rho_f = 0.85\frac{f_c'}{f_{fe}}\left(\frac{a}{d}\right) - \rho_s\frac{f_y}{f_{fe}} = 0.85 \times \frac{5}{117.36} \times \frac{4.46}{16.25} - 0.01083 \times \frac{83.5}{117.36} = 0.00223$$

$$A_f = \rho_f bd = 0.00223 \times 10 \times 16.25 = 0.363 \text{ in}^2$$

$$A_{strip} = 0.63 \times 0.079 = 0.0498 \text{ in}^2$$

$$\# \text{ of strips} = \frac{A_f}{A_{strip}} = 7.3 \text{ strips}$$

Use eight strips in four grooves with 2-in. spacing between the grooves. According to ACI 440.2R-08, Figure 13.4, the groove dimensions are 3 × 2 × 0.079 = 0.474" by 1.5 × 0.63 = 0.95". However, the same number of CFRP strips was successfully used in a beam with cut grooves of 0.25" × 0.75", as seen in Figure 5.7 (Rasheed et al. 2010).

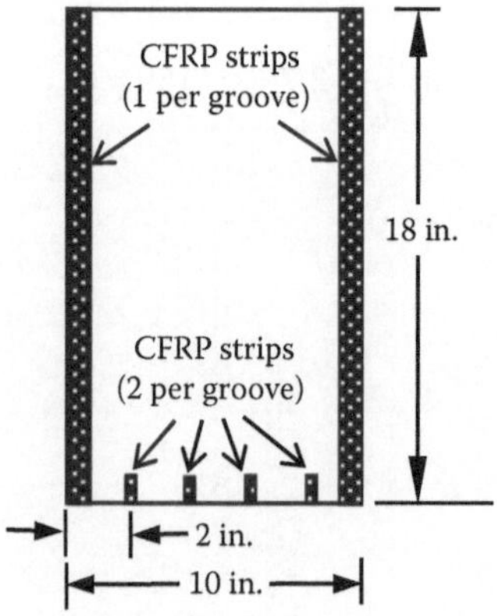

FIGURE 5.7 Example 5.5 showing the CFRP NSM design details.

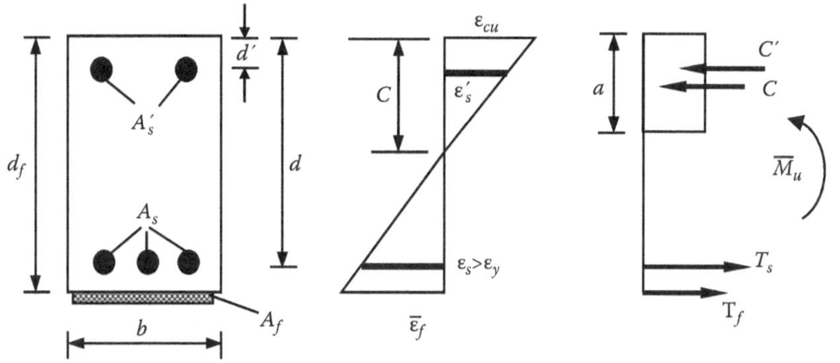

FIGURE 5.8 Doubly reinforced cross section with strain distribution and force profile.

5.4.1.2 Flexural Strengthening of a Doubly Reinforced Section

For the ductile concrete crushing failure mode, ε_{cu} of 0.003 is reached after the yielding of tension reinforcement. The effective strain in the FRP is given by Equation (10.3) of ACI 440.2R-08:

$$\varepsilon_{fe} = 0.003\frac{d_f - c}{c} - \varepsilon_{bi} \le \varepsilon_{fd} \tag{5.21}$$

See Figure 5.8.
From strain compatibility:

$$\varepsilon_s' = 0.003\frac{c - d'}{c} \tag{5.22}$$

Invoking force equilibrium:
$C + C' = T_s + T_f$

$$0.85 f_c' b \beta_1 c + A_s' f_s' = A_s f_y + A_f f_{fe} \tag{5.23}$$

$$\rho_f = 0.85\frac{f_c'}{f_{fe}}\left(\frac{a}{d}\right) - \rho_s\frac{f_y}{f_{fe}} + \rho_s'\frac{f_s'}{f_{fe}} = (\rho_f)_{singly} + \rho_s'\frac{f_s'}{f_{fe}} \tag{5.24}$$

Using moment equilibrium:

$$\bar{M}_u = T_s jd_s + T_f jd_f + C' jd_s'$$

$$\bar{M}_u = \bar{\phi} A_s f_y\left(d - \frac{\beta_1 c}{2}\right) + \bar{\phi}\Psi_f A_f f_{fe}\left(d_f - \frac{\beta_1 c}{2}\right) + \bar{\phi} A_s' f_s'\left(\frac{\beta_1 c}{2} - d'\right) \tag{5.25}$$

Dividing Equation (5.25) by $\bar{\phi} f_c' b d_f^2$,

$$\frac{\bar{M}_u}{\bar{\phi} f_c' b d_f^2} = \rho_s\frac{f_y}{f_c'}\frac{d}{d_f^2}\left(d - \frac{a}{2}\right) + \Psi_f \rho_f\frac{f_{fe}}{f_c'}\frac{d}{d_f^2}\left(d_f - \frac{a}{2}\right) + \rho_s'\frac{f_s'}{f_c'}\frac{d}{d_f^2}\left(\frac{a}{2} - d'\right) \tag{5.26}$$

Substituting Equation (5.24) into Equation (5.26) and rearranging,

$$\left(\frac{a}{d_f}\right)^2 - \left[2 - 0.208(Q_1 - Q_1')\right]\frac{a}{d_f} + 2.77Q_2' = 0 \tag{5.27}$$

where

$$Q_1 = \rho_s \frac{f_y}{f_c'} \frac{d}{d_f} \tag{5.28}$$

$$Q_1' = \rho_s' \frac{f_s'}{f_c'} \frac{d}{d_f} \tag{5.29}$$

$$Q_2' = Q_2 - Q_1'\left(\Psi_f - \frac{d'}{d_f}\right) \quad \text{and} \quad Q_2 \text{ from Equation (5.16)} \tag{5.30}$$

$$\text{Assuming } B \approx 2 \Rightarrow \frac{a}{d_f} = 1 - \sqrt{1 - 2.77Q_2'} \tag{5.31}$$

Q_2' in Equation (5.30) may be directly applied to find $\frac{a}{d_f}$ if compression steel has yielded (which is typically assumed and then checked using the strain compatibility, Equation [5.22]). If this yielding happens, the solution is complete. Otherwise, Equation (5.22) is substituted into Equations (5.29) and (5.30), resulting in a cubic expression,

$$\left(\frac{a}{d_f}\right)^3 - 2\left(\frac{a}{d_f}\right)^2 + 2.77Q_2''\left(\frac{a}{d_f}\right) + 2.77(Q_2 - Q_2'')\frac{\beta_1 d'}{d_f} = 0 \tag{5.32}$$

where

$$Q_2'' = Q_2 - \rho_s' \frac{87}{f_c'} \frac{d}{d_f}\left(\Psi_f - \frac{d'}{d_f}\right) \quad \text{when U.S. customary units are used} \tag{5.33a}$$

$$Q_2'' = Q_2 - \rho_s' \frac{600}{f_c'} \frac{d}{d_f}\left(\Psi_f - \frac{d'}{d_f}\right) \quad \text{when S.I. units are used} \tag{5.33b}$$

Similar equations were first derived by Rasheed and Pervaiz (2003) for doubly reinforced sections with $\Psi_f = 1.0$.

Example 5.6: Analysis

Redo Example 5.3 assuming a doubly reinforced concrete section and considering $f_y' = 83.5$ ksi as well.

Solution:

$$A_s' = 2\times 0.11 = 0.22\text{ in.}^2$$

$$\varepsilon_y' = \frac{f_y'}{E_s} = \frac{83.5}{29000} = 0.00288$$

$$d' = 1'' + \frac{3}{8} + \frac{1}{2}\times\frac{3}{8} = 1.563''$$

Assume ductile crushing failure mode:

$$0.85f_c'b\beta_1 c + A_s'f_s' = A_sf_y + A_ff_{fe}$$

Assume yielding of compression steel:

$$0.85\times 5\times 0.8\times 10c + 0.22\times 83.5 = 1.76\times 83.5 + 0.286\times 99\left(\frac{16.36-c}{c}\right)$$

$$34c^2 + 18.37c - 146.96c - 463.22 + 28.31c = 0$$

$$34c^2 - 100.28c - 463.22 = 0$$

$$c = 5.45\text{ in}$$

$$\varepsilon_s' = \frac{0.003}{5.45}\times(5.45-1.563) = 0.00214 < 0.00288 \Rightarrow \text{Compression steel does not yield}$$

Redo the force equilibrium accordingly:

$$34c + 0.22\times 87\left(\frac{c-1.563}{c}\right) = 146.96 + 28.31\left(\frac{16.36-c}{c}\right)$$

$$34c^2 + 19.14c - 29.92 = 146.96c + 463.22 - 28.31c$$

$$34c^2 - 99.51c - 493.14 = 0$$

$$c = 5.54\text{ in}$$

$$\varepsilon_{fe} = 0.003\frac{16.36-5.54}{5.54} = 0.00586 < \varepsilon_{fd} = 0.00896 < 0.9C_e\varepsilon_{fu} = 0.01197$$

$$f_{fe} = 33{,}000\times 0.00586 = 193.38\text{ ksi}$$

$$\varepsilon_s' = 0.003\frac{5.54-1.563}{5.54} = 0.002154$$

$$f_s' = 62.45\text{ ksi}$$

$$\bar{M}_n = A_sf_y\left(d - \frac{\beta_1 c}{2}\right) + \Psi_f A_f f_{fe}\left(d_f - \frac{\beta_1 c}{2}\right) + A_s'f_s'\left(\frac{\beta_1 c}{2} - d'\right)$$

$$\bar{M}_n = 1.76\times 83.5\times\left(16.25 - \frac{0.8\times 5.54}{2}\right) + 0.85\times 0.286\times 193.38\times\left(16.36 - \frac{0.8\times 5.54}{2}\right)$$

$$+0.22\times 62.45\times\left(\frac{0.8\times 5.54}{2} - 1.563\right) = 2736.33\text{ k-in} = 228.03\text{ k-ft}$$

Compare to 247.6 k-ft from experiment.

Why is there a difference? The actual $\Psi_f = 1.0$, and the small strain hardening in tension steel is ignored here. Considering $\Psi_f = 1.0$,

$$\bar{M}_n = 1.76 \times 83.5 \times \left(16.25 - \frac{0.8 \times 5.54}{2}\right) + 1.0 \times 0.286 \times 193.38 \times \left(16.36 - \frac{0.8 \times 5.54}{2}\right)$$

$$+ 0.22 \times 62.45 \times \left(\frac{0.8 \times 5.54}{2} - 1.563\right) = 2853.67 \text{ k-in} = 237.81 \text{ k-ft}$$

The small difference that still exists is due to some strain hardening in the tension steel (Rasheed et al. 2010).

Example 5.7: Design

Chaallal, Nollet, and Perraton (1998) presented a CFRP flexural strengthening design example of a doubly reinforced simple beam (Figure 5.9). They designed the beam using an iterative approach. The reinforced concrete section details and material properties are shown below.

Solution:

Assume ductile crushing failure mode
Assume yielding of compression steel at failure
Assume $d_f = h = 600$ mm

$$Q_1 = \rho_s \frac{f_y}{f_c'} \frac{d}{d_f} = \frac{2400}{350 \times 535} \times \frac{400}{30} \times \frac{535}{600} = 0.1524$$

$$Q_1' = \rho_s' \frac{f_s'}{f_c'} \frac{d}{d_f} = \frac{400}{350 \times 535} \times \frac{400}{30} \times \frac{535}{600} = 0.0254$$

$$Q_2' = \frac{\bar{M}_u}{\phi f_c' b d_f^2} + Q_1\left(\Psi_f - \frac{d}{d_f}\right) - Q_1'\left(\Psi_f - \frac{d'}{d_f}\right)$$

$$= \frac{500 \times 10^6}{0.9 \times 30 \times 350 \times 600^2} + 0.1524\left(0.85 - \frac{535}{600}\right) - 0.0254\left(0.85 - \frac{58}{600}\right) = 0.1215$$

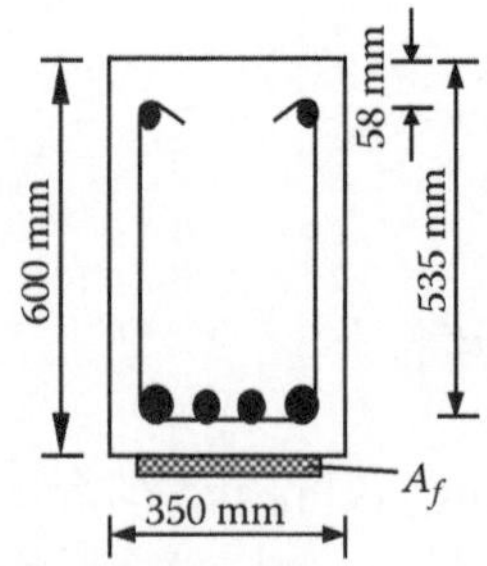

Concrete Properties	Steel Properties	CFRP Properties
$f_c' = 30$ MPa	$f_y = 400$ MPa	$f_{fu} = 2400$ MPa
$E_c = 24647$ MPa	$E_s = 200$ GPa	$E_f = 150$ GPa
$\varepsilon_{cu} = 0.003$	$A_s = 2400$ mm^2	ε_{bi} = Zero
$\beta_1 = 0.833$	$A_s' = 400$ mm^2	$t_f = 0.17$ mm
M_u = 380 KN.m		
$\bar{M}_u$ = 500 KN.m		

FIGURE 5.9 Design of a doubly reinforced beam example by Chaallal, Nollet, and Perraton (1998).

$$\left(\frac{a}{d_f}\right)^2 - 1.974\left(\frac{a}{d_f}\right) + 2.77 \times 0.1215 = 0$$

$$\frac{a}{d_f} = \frac{1.974 - \sqrt{1.974^2 - 4 \times 2.77 \times 0.1215}}{2} = 0.1887$$

$$a = 113.25 \text{ mm} \Rightarrow c = 135.95 \text{ mm}$$

$$\varepsilon_s' = 0.003\frac{c - d'}{c} = 1.72 \times 10^{-3} < \frac{400}{200{,}000} = 2 \times 10^{-3} \Rightarrow \text{ No yielding N.G.}$$

$$Q_2 = \frac{\bar{M}_u}{\bar{\phi} f_c' b d_f^2} + Q_1\left(\Psi_f - \frac{d}{d_f}\right) = \frac{500 \times 10^6}{0.9 \times 30 \times 350 \times 600^2} + 0.1524\left(0.85 - \frac{535}{600}\right)$$
$$= 0.1406$$

$$Q_2'' = Q_2 - \rho_s' \frac{600}{f_c'} \frac{d}{d_f}\left(\Psi_f - \frac{d'}{d_f}\right) = 0.1406 - \frac{400}{350 \times 535} \times \frac{600}{30} \times \frac{535}{600}\left(0.85 - \frac{58}{600}\right)$$
$$= 0.1119$$

$$\left(\frac{a}{d_f}\right)^3 - 2\left(\frac{a}{d_f}\right)^2 + 2.77 \times 0.1119\left(\frac{a}{d_f}\right) + 2.77(0.1406 - 0.1119) \times \frac{0.833 \times 58}{600} = 0$$

$$\left(\frac{a}{d_f}\right)^3 - 2\left(\frac{a}{d_f}\right)^2 + 0.31\left(\frac{a}{d_f}\right) + 6.4 \times 10^{-3} = 0$$

$$\frac{a}{d_f} = 0.1893 \rightarrow \text{ By finding the smallest positive root using Excel Goal Seek}$$

$$a = 113.58 \text{ mm} \Rightarrow c = \frac{113.58}{0.833} = 136.35 \text{ mm}$$

$$\varepsilon_s' = 0.003\frac{136.35 - 58}{136.35} = 1.724 \times 10^{-3}$$

$$f_s' = 200000 \times 1.724 \times 10^{-3} = 344.78 \text{ MPa}$$

$$\varepsilon_{fe} = 0.003\frac{600 - 136.35}{136.35} - 0 = 0.0102 \approx 0.009944$$

$$\varepsilon_{fd} = 0.41\sqrt{\frac{30}{2 \times 150{,}000 \times 0.17}} = 0.009944$$

$$f_{fe} = 150{,}000 \times 0.0102 = 1{,}530 \text{ MPa}$$

$$\varepsilon_s = 0.003\frac{535 - 136.35}{136.35} = 0.00877 > 0.005$$

so $\bar{\phi} = 0.9$

$$\rho_f = 0.85\frac{f_c'}{f_{fe}}\left(\frac{a}{d}\right) - \rho_s\frac{f_y}{f_{fe}} + \rho_s'\frac{f_s'}{f_{fe}}$$

$$= 0.85 \times \frac{30}{1530} \times \frac{113.58}{535} - \frac{2400}{350 \times 535} \times \frac{400}{1530} + \frac{400}{350 \times 535} \times \frac{344.78}{1530} = 6.688 \times 10^{-4}$$

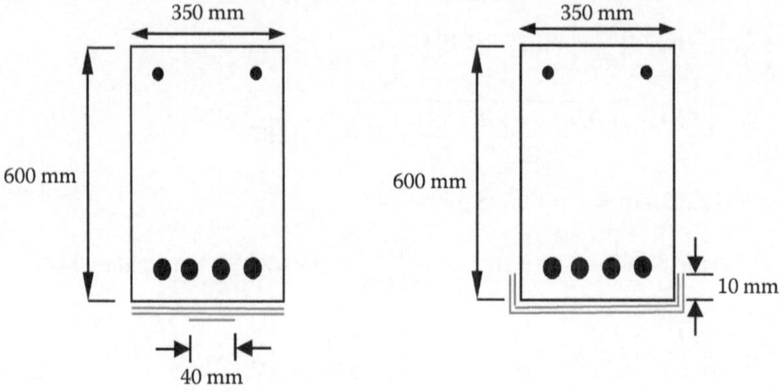

FIGURE 5.10 Example 5.7 showing the alternative FRP design details.

$$A_f = \rho_f \times bd = 125.23 \text{ mm}^2$$

$$b_f = \frac{A_f}{nt_f} = \frac{125.23}{2 \times 0.17} = 368.32 \text{ mm} > 350 \text{ mm}$$

$$2 \times 0.17 \times 350 + b_f' \times 0.17 = 125.23$$

$$b_f' = 36.7 \text{ mm} \approx 40\text{mm}$$

Either wrap the two layers 10 mm up the sides or add a 40-mm partial layer to the soffit, as shown in Figure 5.10.

The effective depth of the wrapped CFRP layers in the second alternative is

$$d_f = \frac{600.17 \times 350 \times 0.34 + 2 \times 595 \times 10 \times 0.34}{350 \times 0.34 + 2 \times 10 \times 0.34} = 599.89 \approx 600 \text{ mm O.K.}$$

5.4.2 Brittle Crushing of Concrete

This is the failure mode in which concrete reaches the limit of useful compressive strain of 0.003 before yielding of tension reinforcement (i.e., $\varepsilon_s < \varepsilon_y$). It is not a desirable failure mode because it does not involve a warning sign at failure. Therefore, ACI 440.2R-08 reduces the ϕ factor to 0.65 to make this failure mode less likely to occur. It is not an anticipated failure mode, but it is allowed by ACI 440.2R-08. This section will show that this failure mode, which is admissible for a heavily reinforced, heavily strengthened section, is not practical to consider.

5.4.2.1 Flexural Strengthening of a Singly Reinforced Section

Strain compatibility:

$$\varepsilon_s = 0.003 \frac{d-c}{c} \tag{5.34}$$

$$\varepsilon_{fe} = 0.003 \frac{d_f - c}{c} - \varepsilon_{bi} \le \varepsilon_{fd} \tag{5.35}$$

Force equilibrium:

$$0.85 f_c' b \beta_1 c = A_s f_s + A_f f_{fe} \tag{5.36}$$

$$A_f = \frac{0.85 f_c' b a - A_s f_s}{f_{fe}} \tag{5.37}$$

$$\rho_f = 0.85 \frac{f_c'}{f_{fe}} \frac{a}{d} - \rho_s \frac{f_s}{f_{fe}} \tag{5.38}$$

Moment equilibrium:

$$\bar{M}_u = \bar{\phi}\bar{M}_n = \bar{\phi} A_s f_s \left(d - \frac{a}{2} \right) + \bar{\phi} \Psi_f A_f f_{fe} \left(d_f - \frac{a}{2} \right) \tag{5.39}$$

$$\frac{\bar{M}_u}{\bar{\phi} f_c' b d_f^2} = \rho_s \frac{f_s}{f_c'} \frac{d}{d_f^2} \left(d - \frac{a}{2} \right) - \Psi_f \rho_f \frac{f_{fe}}{f_c'} \frac{d}{d_f^2} \left(d_f - \frac{a}{2} \right) \tag{5.40}$$

Substituting Equation (5.38) into Equation (5.40) and rearranging the terms,

$$\left(\frac{a}{d_f} \right)^3 - \left(2 + 0.208 \hat{Q}_1 \right) \left(\frac{a}{d_f} \right)^2 + \left(2.77 Q_2 + 0.208 \hat{Q}_1 \beta_1 \frac{d}{d_f} \right) \left(\frac{a}{d_f} \right)$$

$$+ 2.77 \hat{Q}_1 \frac{\beta_1 d}{d_f} \left(\Psi_f - \frac{d}{d_f} \right) = 0 \tag{5.41}$$

where

$$\hat{Q}_1 = \rho_s \frac{87}{f_c'} \frac{d}{d_f} \quad \text{when U.S. customary units are used} \tag{5.42a}$$

$$\hat{Q}_1 = \rho_s \frac{600}{f_c'} \frac{d}{d_f} \quad \text{when S.I. units are used} \tag{5.42b}$$

$$Q_2 = \frac{\bar{M}_u}{\bar{\phi} f_c' b d_f^2} - \hat{Q}_1 \left(\Psi_f - \frac{d}{d_f} \right) \tag{5.43}$$

Now we need to determine at which level of FRP and steel reinforcement the mode of failure switches from ductile crushing to brittle crushing. For that, we need to derive a balanced FRP ratio at which concrete extreme-compression fiber reaches 0.003 at the same time that $\varepsilon_s = \varepsilon_y$. This is called ρ_f^{bal}, and it is derived by Rasheed and Pervais (2003) as follows:

$$\frac{\varepsilon_{cu}}{c^{bal}} = \frac{\varepsilon_{cu} + \varepsilon_y}{d} \Rightarrow c^{bal} = \frac{0.003}{0.003 + \varepsilon_y} d \tag{5.44}$$

$$c^{bal} = \frac{87}{87 + f_y} d \text{ and } f_y \text{ is in ksi when U.S. customary units are used} \tag{5.45a}$$

$$c^{bal} = \frac{600}{600 + f_y} d \text{ and } f_y \text{ is in MPa when S.I. units are used} \tag{5.45b}$$

$$\frac{\varepsilon_{cu}}{c^{bal}} = \frac{\varepsilon_{cu} + \bar{\varepsilon}_f^{bal}}{d_f} \Rightarrow \varepsilon_f^{bal} = \left(\frac{d_f}{c^{bal}} - 1 \right) \varepsilon_{cu} - \varepsilon_{bi} \tag{5.46}$$

$$\varepsilon_f^{bal} = 0.003 \left(\frac{87 + f_y}{87} \frac{d_f}{d} - 1 \right) - \varepsilon_{bi} \text{ and } f_y \text{ is in ksi when U.S. customary units are used} \tag{5.47a}$$

$$\varepsilon_f^{bal} = 0.003 \left(\frac{600 + f_y}{600} \frac{d_f}{d} - 1 \right) - \varepsilon_{bi} \text{ and } f_y \text{ is in MPa when S.I. units are used} \tag{5.47b}$$

From force equilibrium:

$$0.85 f_c' b \beta_1 c^{bal} = A_s f_y + A_f^{bal} E_f \varepsilon_f^{bal}$$

Dividing both sides by ($f_y bd$) results in

$$\frac{E_f \varepsilon_f^{bal}}{f_y} \rho_f^{bal} = 0.85 \beta_1 \frac{f_c'}{f_y} \frac{87}{87 + f_y} - \rho_s \text{ and } f_y \text{ is in ksi when U.S. customary units are used} \tag{5.48a}$$

$$\frac{E_f \varepsilon_f^{bal}}{f_y} \rho_f^{bal} = 0.85 \beta_1 \frac{f_c'}{f_y} \frac{600}{600 + f_y} - \rho_s \text{ and } f_y \text{ is in MPa when S.I. units are used} \tag{5.48b}$$

Therefore

$$\rho_f^{bal} = \frac{f_y}{f_f^{bal}} \left(\rho_s^{bal} - \rho_s \right) \tag{5.48c}$$

where

$$\rho_f^{bal} = \frac{A_f^{bal}}{bd} \tag{5.49}$$

$$\rho_s = \frac{A_s}{bd} \tag{5.50}$$

$$\rho_s^{bal} = 0.85 \beta_1 \frac{f_c'}{f_y} \frac{87}{87 + f_y} \text{ and } f_y \text{ is in ksi when U.S. customary units are used} \tag{5.51a}$$

$$\rho_s^{bal} = 0.85\beta_1 \frac{f_c'}{f_y}\frac{600}{600+f_y} \quad \text{and } f_y \text{ is in MPa when S.I. units are used} \qquad (5.51b)$$

It is evident from Equation (5.48) that the more heavily reinforced the section is with tension steel reinforcement, the less likely it is for the ductile crushing failure mode to be admitted and the higher is the chance of the brittle failure mode to take place.

Example 5.8: Analysis

The beam section shown in Figure 5.11 is heavily reinforced with steel bars. Determine the unstrengthened and strengthened moment capacity of this section.

$f_c' = 4$ ksi $\qquad f_y = 60$ ksi $\qquad t_f = 0.04''$ $\qquad$ Clear cover = 1″

Three layers of Tyfo CFRP wrapped 2" around the sides (Figure 5.11)

$E_f = 13{,}900$ ksi $\qquad f_{fu} = 143$ ksi $\qquad \varepsilon_{bi} = 0.0008$

Solution:

First check the failure mode likely to occur:

$$d = d_t = 12'' - 1.0 - \frac{3}{8} - \frac{1}{2}\times\frac{7}{8} = 10.19''$$

$$c^{bal} = \frac{87}{87+f_y}d = \frac{87}{87+60}\times 10.19 = 6.03''$$

$$d_f = \frac{3\times16\times0.04\times12 + 3\times2\times2\times0.04\times11}{3\times16\times0.04 + 3\times2\times2\times0.04} = 11.8 \text{ in}$$

$$\varepsilon_f^{bal} = 0.003\left(\frac{87+f_y}{87}\frac{d_f}{d} - 1\right) - \varepsilon_{bi}$$

$$\varepsilon_f^{bal} = 0.003\left(\frac{d_f}{c^{bal}} - 1\right) - \varepsilon_{bi} = 0.003\left(\frac{11.8}{6.03} - 1\right) - 0.0008 = 0.00207$$

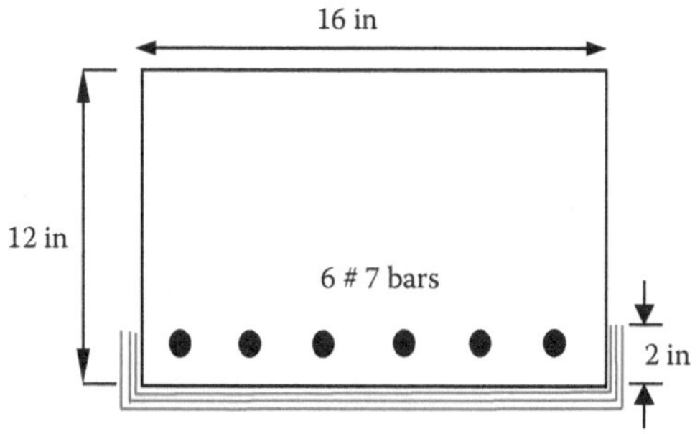

FIGURE 5.11 Example 5.8 showing the section details and the FRP suggested area.

$$f_f^{bal} = E_f \varepsilon_f^{bal} = 13{,}900 \times 0.00207 = 28.77 \text{ ksi}$$

$$A_s = 6 \times 0.6 = 3.6 \; in^2$$

$$b = 2\times1 + 2\times\frac{3}{8} + 6\times0.875 + 5\times1 = 13'' < 16'' \quad \text{O.K.}$$

$$\rho_s = \frac{A_s}{bd} = \frac{3.6}{16\times10.19} = 0.02208$$

$$\rho_s^{bal} = 0.85\beta_1 \frac{f_c'}{f_y}\frac{87}{87+f_y} = 0.85\times0.85\times\frac{4}{60}\times\frac{87}{87+60} = 0.02851$$

$$\rho_f^{bal} = \frac{f_y}{f_f^{bal}}\left(\rho_s^{bal} - \rho_s\right) = \frac{60}{28.77}(0.02851 - 0.02208) = 0.0134$$

$$A_f^{bal} = \rho_f^{bal} \times bd = 0.0134\times16\times10.19 = 2.186 \text{ in}^2$$

$$A_f = 3\times(16+4)\times0.04 = 2.4 \text{ in}^2$$

$$A_f > A_f^{bal} \Rightarrow \text{brittle crushing failure}$$

Unstrengthened beam analysis:
Assuming that steel yields:

$$c = \frac{A_s f_y}{0.85 f_c' b \beta_1} = \frac{3.6\times60}{0.85\times4\times16\times0.85} = 4.67''$$

$$a = 0.85\times4.67'' = 3.97''$$

$$\varepsilon_s = 0.003\times\frac{d-c}{c} = 0.003\times\frac{10.19-4.67}{4.67} = 0.00355 < 0.004 \quad \text{N.G.}$$

$$\phi = 0.65 + (0.00355 - 0.002)\frac{250}{3} = 0.779$$

or

$$\frac{c}{d_t} = \frac{4.67}{10.19} = 0.458 > 0.429 \quad \text{N.G.}$$

$$\phi = 0.65 + 0.25\left(\frac{1}{0.458} - \frac{5}{3}\right) = 0.779$$

Even though the beam is slightly not up to ACI 318-11 (2011) provisions, continue the analysis, since it is an existing beam and it is allowed to be strengthened by ACI 440.2R-08.

$$M_u = \phi A_s f_y\left(d - \frac{a}{2}\right) = 0.779\times3.6\times60\times\left(10.19 - \frac{3.97}{2}\right) = 1380.6 \text{ k-in.}$$

Strengthened beam analysis:
It is known that the failure mode is brittle crushing (prior to yielding of steel), so

$$0.85f_c'b\beta_1c = A_sf_s + A_ff_{fe}$$

$$0.85f_c'b\beta_1c = A_sE_s\left(0.003\frac{d-c}{c}\right) + A_fE_f\left(0.003\frac{d_f-c}{c} - \varepsilon_{bi}\right)$$

$$46.24c^2 = 313.2(10.19-c)+100.08(11.8-c)-26.69c$$

$$46.24c^2 + 439.97c - 4372.45 = 0$$

$$c = \frac{-439.97+\sqrt{439.97^2+4\times 46.24\times 4372.45}}{2\times 46.24} = 6.07'' \Rightarrow a = 5.16''$$

$$\varepsilon_s = \frac{0.003}{c}(d-c) = 0.00204 < \varepsilon_y = 0.00207$$

$$\varepsilon_{fe} = \frac{0.003}{c}(d_f-c) - \varepsilon_{bi} = 0.00203 < \varepsilon_{fd}$$

$$\varepsilon_{fd} = 0.083\sqrt{\frac{4000}{3\times 13900000\times 0.04}} = 0.00406$$

Thus, the brittle crushing failure is confirmed.

$$\bar{M}_u = \bar{\phi}\bar{M}_n = \bar{\phi}A_sf_s\left(d-\frac{a}{2}\right) + \bar{\phi}\Psi_fA_ff_{fe}\left(d_f-\frac{a}{2}\right)$$

$$= 0.65\times 3.6\times 29{,}000\times 0.00204\times\left(10.19-\frac{5.16}{2}\right)+0.65\times 0.85\times 2.4$$

$$\times 13{,}900\times 0.00203\times\left(11.8-\frac{5.16}{2}\right) = 1053.62+345.01 = 1398.63 \text{ k-in.}$$

So for a large amount of CFRP, the section is strengthened by only 1.3%. This example is intended to show the ineffectiveness of this failure mode that is not admissible unless the section is already heavily reinforced to start with. To reinforce this idea even further, the following design example (Example 5.9) is presented.

Example 5.9: Design

For the section analyzed in Example 5.8, design for the required area of FRP A_f and determine the failure mode involved using the ACI 440.2R-08 approach if

$$M_u = 1380.60k - in$$

$$\bar{M}_u = 1518.66k - in$$

(i.e., 10% strengthening level)
Use $d_f = h = 12''$

Solution:

If ρ_f is unknown and Example 5.8 does not suggest a failure mode, assume ductile crushing failure.

$$Q_1 = 0.0221 \times \frac{60}{4} \times \frac{10.19}{12} = 0.2185$$

$$Q_2 = \frac{1518.66}{0.9 \times 4 \times 16 \times 12^2} + 0.2185 \times \left(0.85 - \frac{10.19}{12}\right) = 0.1833$$

$$\frac{a}{d_f} = 1 - \sqrt{1 - 2.77 \times 0.1833} = 0.298$$

$$a = 3.58'' \Rightarrow c = 4.21''$$

$$\varepsilon_s = 0.003\frac{d-c}{c} = 4.256 \times 10^{-3} \text{ ductile transition}$$

$$\frac{c}{d_t} = \frac{4.21''}{10.19''} = 0.413$$

$$\phi = 0.65 + 0.25\left[\frac{1}{0.413} - \frac{5}{3}\right] = 0.838$$

Repeat the $Q_2, \frac{a}{d_f}, c, \varepsilon_s$ calculations, since ϕ has changed. Use Excel to repeat the mechanical iterations.

Q2	a/df	a	c	ε_s	c/dt	Phi
0.183329	0.298446	3.581351	4.213354	0.004256	0.413479	0.837958
0.196885	0.325739	3.90887	4.598671	0.003648	0.451293	0.787298
0.209539	0.352253	4.227037	4.972985	0.003147	0.488026	0.745601
0.221244	0.377784	4.533405	5.333417	0.002732	0.523397	0.710982
0.232006	0.402217	4.826609	5.678364	0.002384	0.557249	0.681966
0.241867	0.425519	5.10623	6.007329	0.002089	0.589532	0.657399
0.250897	0.447718	5.372618	6.320727	0.001836	0.620287	0.650

It is evident that $\phi = 0.65$ and the mode of failure is brittle crushing (compression controlled).

$$\hat{Q}_1 = \rho_s \frac{87}{f_c'}\frac{d}{d_f} = 0.0221 \times \frac{87}{4} \times \frac{10.19}{12} = 0.4082$$

$$Q_2 = \frac{\bar{M}_u}{\phi f_c' b d_f^2} - \hat{Q}_1\left(\Psi_f - \frac{d}{d_f}\right) = \frac{1518.66}{0.65 \times 4 \times 16 \times 12^2} - 0.4082 \times \left(0.85 - \frac{10.19}{12}\right) = 0.2532$$

$$\left(\frac{a}{d_f}\right)^3 - (2 + 0.208 \times 0.4082)\left(\frac{a}{d_f}\right)^2$$
$$+\left(2.77 \times 0.2532 + 0.208 \times 0.4082 \times 0.85 \times \frac{10.19}{12}\right)\left(\frac{a}{d_f}\right)$$
$$+2.77 \times 0.4082 \times 0.85 \times \frac{10.19}{12} \times \left(0.85 - \frac{10.19}{12}\right) = 0$$

$$\left(\frac{a}{d_f}\right)^3 - 2.085\left(\frac{a}{d_f}\right)^2 + 0.7626\left(\frac{a}{d_f}\right) + 0.00068 = 0$$

Solving for the lowest positive root using Excel Goal-Seek function,

$$\frac{a}{d_f} = 0.47425 \Rightarrow a = 5.69'' \Rightarrow c = 6.70''$$

$$\varepsilon_s = 0.003\frac{10.19-6.7}{6.7} = 0.00156 < \varepsilon_y = 0.00207$$

$$\varepsilon_{fe} = 0.003\frac{12-6.7}{6.7} - 0.0008 = 0.00157 < \varepsilon_{fd}$$

$$\varepsilon_{fd} = 0.083\sqrt{\frac{4{,}000}{3\times 13{,}900{,}000\times 0.04}} = 0.00406$$

Thus, brittle crushing failure is confirmed.

$$f_{fe} = 13{,}900\times 0.00157 = 21.823 \text{ ksi}$$

$$\rho_f = 0.85\frac{f_c'}{f_{fe}}\frac{a}{d} - \rho_s\frac{f_s}{f_{fe}} = 0.85\times\frac{4}{21.823}\times\frac{5.69}{10.19} - 0.0221\times\frac{29{,}000\times 0.00156}{21.823}$$

$$= 0.0412 > \rho_f^{bal} = 0.0134 \text{ (Example 5.8)}$$

$$A_f = \rho_f bd = 0.0412\times 16\times 10.19 = 6.714 \text{ in.}^2$$

$$b_f = \frac{A_f}{nt_f} = \frac{6.714}{3\times 0.04} = 55.95'' >> 16 + 2\times\frac{12}{2} = 28'' \text{ assuming NA @ } \frac{h}{2}$$

This is an impractical strengthening case. The section that is heavily reinforced cannot be strengthened by 10%.

5.4.3 Rupture of FRP

This is one of the ductile flexural failure modes of FRP-strengthened beams, since the internal steel reinforcement is guaranteed to yield way prior to the rupture of FRP. It is a feature of lightly reinforced and lightly strengthened sections, which typically happens in slabs and T-beams. Despite the ductile nature of this failure mode, FRP rupture is sudden and catastrophic, since it is typically accompanied by a significant release of elastic energy. Accordingly, ACI 440.2R-08 tries to lessen the effect of this failure mode by limiting the ultimate FRP strain allowed to 90% of the ε_{fu}. This is in addition to the environmental factor C_E that multiplies the design ultimate strain. The latter is known to be the mean tensile strain value minus three times the standard deviation to guarantee a 99.87% probability of exceedance, as shown in Chapter 3.

5.4.3.1 Maximum FRP Reinforcement Ratio for Rupture Failure Mode

To ensure that this mode controls the design, the FRP ratio should be kept below the balanced ratio that would cause simultaneous ductile concrete crushing and the FRP rupture limit (0.9 ε_{fu}). This ratio is determined using expressions developed by Rasheed and Pervaiz (2003) for singly and doubly reinforced rectangular sections:

$$\rho_f^{b\max} = \frac{A_f^{b\max}}{bd} = 0.85\frac{f_c'}{0.9f_{fu}}\frac{a_b^{\max}}{d} - \rho_s\frac{f_y}{0.9f_{fu}} \tag{5.52}$$

where

$$a_b^{\max} = \beta_1 c_b^{\max} = \beta_1\frac{\varepsilon_{cu}\, d_f}{\varepsilon_{cu} + \bar{\varepsilon}_{fu}^{\max}} \tag{5.53}$$

$$\bar{\varepsilon}_{fu}^{\max} = 0.9\varepsilon_{fu} + \varepsilon_{bi} \tag{5.54}$$

$$\rho_s = \frac{A_s}{bd} \tag{5.55}$$

$$\tilde{\rho}_f^{b,\max} = \rho_f^{b,\max} + \rho_s'\frac{f_s'^{b,\max}}{0.9f_{fu}} \tag{5.56}$$

$$f_s'^{b,\max} = f_y \quad \text{if} \quad \frac{d'}{d_f} \le \frac{87 - f_y}{87 + 29{,}000\bar{\varepsilon}_{fu}^{\max}}$$

$$= 87 - \frac{d'}{d_f}\left(87 + 29{,}000\bar{\varepsilon}_{fu}^{\max}\right) \quad \text{if} \quad \frac{d'}{d_f} > \frac{87 - f_y}{87 + 29{,}000\bar{\varepsilon}_{fu}^{\max}} \tag{5.57}$$

Note that Equation (5.57) is used with U.S. customary units (f_y in ksi) and that every (87) in the equation is replaced with (600) for the case of S.I. units (f_y in MPa). Also, every (29,000) is replaced with (200,000) in the case of S.I. units.

5.4.3.2 Exact Solution for Singly Reinforced Rectangular Sections

Force equilibrium:

$$\sum F_x = 0 \Rightarrow \alpha f_c' bc - A_s f_y - A_f\left(0.9f_{fu}\right) = 0 \tag{5.58}$$

$$\rho_f = \frac{A_f}{bd} = \alpha\frac{f_c'}{0.9f_{fu}}\frac{c}{d} - \rho_s\frac{f_y}{0.9f_{fu}} \tag{5.59}$$

Moment equilibrium:

$$\bar{M}_u = \bar{\phi}\bar{M}_n = \bar{\phi}A_s f_y(d - \beta c) + \bar{\phi}\Psi_f A_f\left(0.9f_{fu}\right)\left(d_f - \beta c\right) \tag{5.60}$$

$$\frac{\bar{M}_u}{\bar{\phi}f_c' b d_f^2} = \rho_s\frac{f_y}{f_c'}\frac{d^2}{d_f^2}\left(1 - \frac{\beta c}{d}\right) + \Psi_f\rho_f\frac{\left(0.9f_{fu}\right)}{f_c'}\frac{d}{d_f}\left(1 - \frac{\beta c}{d_f}\right) \tag{5.61}$$

Substituting Equation (5.59) into Equation (5.61),

$$\frac{\bar{M}_u}{\phi f_c' b d_f^2} = Q_1 \frac{d}{d_f}\left(1-\frac{\beta c}{d}\right) + \Psi_f \left[\alpha \frac{f_c'}{0.9 f_{fu}} \frac{c}{d} - \rho_s \frac{f_y}{0.9 f_{fu}}\right] \frac{(0.9 f_{fu})}{f_c'} \frac{d}{d_f}\left(1-\frac{\beta c}{d_f}\right)$$

$$\frac{\bar{M}_u}{\phi f_c' b d_f^2} = Q_1 \frac{d}{d_f}\left(1-\frac{\beta c}{d}\right) + \Psi_f \frac{\alpha c}{d_f}\left(1-\frac{\beta c}{d_f}\right) - \Psi_f Q_1 \left(1-\frac{\beta c}{d_f}\right) \tag{5.62}$$

where

$$Q_1 = \rho_s \frac{f_y}{f_c'} \frac{d}{d_f} \tag{5.63}$$

Recalling the α expression from Equations (2.20) and substituting the strain compatibility expression for ε_{cf},

$$\alpha = \frac{\varepsilon_{cf}}{\varepsilon_c'} - \frac{1}{3}\left(\frac{\varepsilon_{cf}}{\varepsilon_c'}\right)^2 = \frac{\bar{\varepsilon}_{fu}^{\max} c}{(d_f - c)\,\varepsilon_c'} - \frac{1}{3}\left(\frac{\bar{\varepsilon}_{fu}^{\max} c}{(d_f - c)\,\varepsilon_c'}\right)^2 \tag{5.64}$$

Introducing the β expression for the centroid location of the Hognestad's parabola, with respect to the top extreme fiber, and substituting the strain compatibility expression for ε_{cf},

$$\beta = \frac{\frac{1}{3} - \frac{\varepsilon_{cf}}{12\varepsilon_c'}}{1 - \frac{\varepsilon_{cf}}{3\varepsilon_c'}} = \frac{\frac{1}{3} - \frac{\bar{\varepsilon}_{fu}^{\max} c}{12(d_f - c)\,\varepsilon_c'}}{1 - \frac{\bar{\varepsilon}_{fu}^{\max} c}{3(d_f - c)\,\varepsilon_c'}} \tag{5.65}$$

Substituting Equations (5.64) and (5.65) into Equation (5.62) and rearranging the terms,

$$A\frac{c}{d_f} + B\left(\frac{c}{d_f}\right)^2 + D\left(\frac{c}{d_f}\right)^3 + E\left(\frac{c}{d_f}\right)^4 + F\left(\frac{c}{d_f}\right)^5 + 9Q_2\varepsilon_c'^3 = 0 \tag{5.66}$$

where

$$A = -27Q_2\varepsilon_c'^3 - 3Q_2\varepsilon_c'^2\bar{\varepsilon}_{fu}^{\max} + 3Q_1\varepsilon_c'^3\left(1-\Psi_f\right)$$

$$= 3\left(Q_1\left(1-\Psi_f\right) - 9Q_2\right)\varepsilon_c'^3 - 3Q_2\varepsilon_c'^2\bar{\varepsilon}_{fu}^{\max} \tag{5.67}$$

$$B = 27Q_2\varepsilon_c'^3 + 6Q_2\varepsilon_c'^2\overline{\varepsilon}_{fu}^{\max} - 9Q_1\varepsilon_c'^3(1-\Psi_f) - 3\left(\frac{Q_1}{4} + 3\Psi_f\right)\varepsilon_c'^2\overline{\varepsilon}_{fu}^{\max} + \frac{3}{4}\Psi_f Q_1\varepsilon_c'^2\overline{\varepsilon}_{fu}^{\max}$$

$$= 9(3Q_2 - Q_1(1-\Psi_f))\varepsilon_c'^3 + 3\left(2Q_2 - \frac{Q_1}{4}(1-\Psi_f) - 3\Psi_f\right)\varepsilon_c'^2\overline{\varepsilon}_{fu}^{\max} \tag{5.68}$$

$$D = -9Q_2\varepsilon_c'^3 - 3Q_2\varepsilon_c'^2\overline{\varepsilon}_{fu}^{\max} + 9Q_1\varepsilon_c'^3(1-\Psi_f) + 6\left(\frac{Q_1}{4} + 3\Psi_f\right)\varepsilon_c'^2\overline{\varepsilon}_{fu}^{\max} + 6\Psi_f\varepsilon_c'\overline{\varepsilon}_{fu}^{\max 2}$$

$$+ 3\Psi_f\varepsilon_c'^2\overline{\varepsilon}_{fu}^{\max} - \frac{6}{4}Q_1\Psi_f\varepsilon_c'^2\overline{\varepsilon}_{fu}^{\max}$$

$$\mathrm{D} = 9\big(Q_1(1-\Psi_f) - Q_2\big)\varepsilon_c'^3 + 3\left(\frac{Q_1}{2}(1-\Psi_f) + 6\Psi_f - Q_2\right)\varepsilon_c'^2\overline{\varepsilon}_{fu}^{\max} + 6\Psi_f\varepsilon_c'\overline{\varepsilon}_{fu}^{\max 2} \tag{5.69}$$

$$E = -3Q_1(1-\Psi_f)\varepsilon_c'^3 - 3\left(\frac{Q_1}{4} + 3\Psi_f\right)\varepsilon_c'^2\overline{\varepsilon}_{fu}^{\max} - 6\Psi_f\varepsilon_c'\overline{\varepsilon}_{fu}^{\max 2} - \Psi_f\overline{\varepsilon}_{fu}^{\max 3} - 6\Psi_f\varepsilon_c'^2\overline{\varepsilon}_{fu}^{\max}$$

$$-\frac{7}{4}\Psi_f\varepsilon_c'\overline{\varepsilon}_{fu}^{\max 2} + \frac{3}{4}Q_1\Psi_f\varepsilon_c'^2\overline{\varepsilon}_{fu}^{\max}$$

$$E = -3Q_1(1-\Psi_f)\varepsilon_c'^3 - 3\left(\frac{Q_1}{4}(1-\Psi_f) + 5\Psi_f\right)\varepsilon_c'^2\overline{\varepsilon}_{fu}^{\max} - \frac{31}{4}\Psi_f\varepsilon_c'\overline{\varepsilon}_{fu}^{\max 2} - \Psi_f\overline{\varepsilon}_{fu}^{\max 3} \tag{5.70}$$

$$F = 3\Psi_f\varepsilon_c'^2\overline{\varepsilon}_{fu}^{\max} + \frac{7}{4}\Psi_f\varepsilon_c'\overline{\varepsilon}_{fu}^{\max 2} + \frac{1}{4}\Psi_f\overline{\varepsilon}_{fu}^{\max 3} \tag{5.71}$$

This equation is not practical to solve in design. There are two alternatives to use in design. The first one is an approximate solution for $\frac{c}{d_f}$, and the second one is an almost exact statistically correlated linear equation (Rasheed and Motto 2010; Saqan, Rasheed, and Hawileh 2013).

5.4.3.3 Approximate Solution for Singly Reinforced Rectangular Sections

If ε_c' is approximated using the typical value of 0.002, which is accurate for normal strength concrete, the fifth-degree polynomial is reduced to a cubic polynomial (Rasheed and Motto 2010).

The α and β expressions will be extremely simplified by the substitution of $\varepsilon_c' = 0.002$:

$$\alpha = 500\varepsilon_{cf} - 83333\varepsilon_{cf}^2 \tag{5.72}$$

$$\beta = \frac{0.33 - 41.67\varepsilon_{cf}}{1 - 166.67\varepsilon_{cf}} \tag{5.73}$$

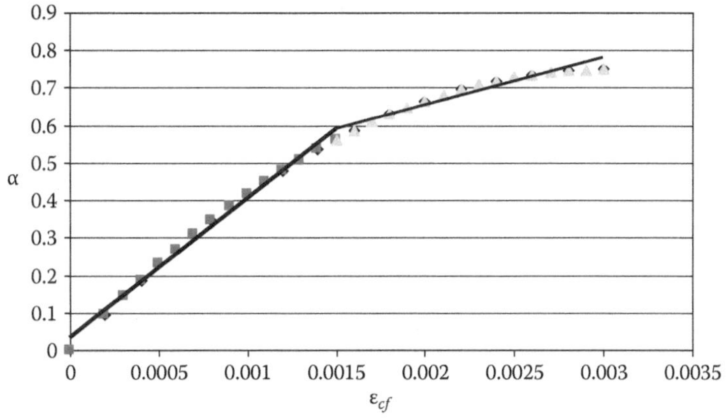

FIGURE 5.12 Linear regression plot of α vs. ε_{cf} relationship. (First published by Engineers Australia. Reprinted with permission.)

Plotting α and β functions in terms of ε_{cf}, one notices that α may be approximated by two straight lines with a breaking point at around $\varepsilon_{cf} = 0.0015$ with an $R^2 = 0.9931$ and 0.9305, respectively, as seen in Figure 5.12.

$$\alpha = 366.67\varepsilon_{cf} + 0.0417 \qquad 0 \le \varepsilon_{cf} < 0.0015$$

$$\alpha = 125\varepsilon_{cf} + 0.4042 \qquad 0.0015 \le \varepsilon_{cf} \le 0.003 \qquad (5.74)$$

In addition, β is seen to have a slight variation along the entire range of ε_{cf}, as seen in Figure 5.13:

$$\beta = 27.768\varepsilon_{cf} + 0.3239 \qquad (R^2 = 0.9677) \qquad (5.75)$$

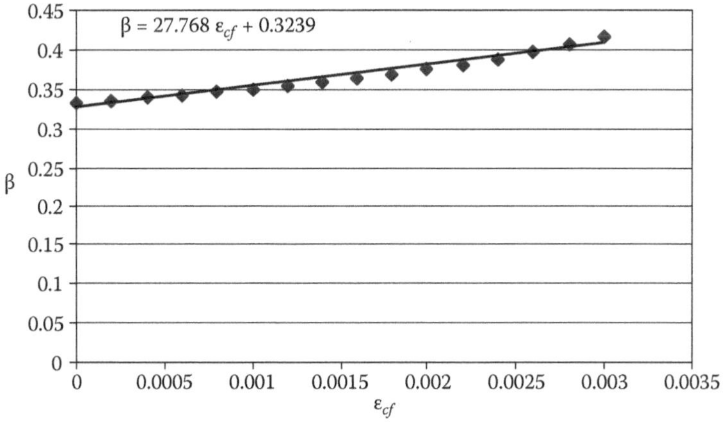

FIGURE 5.13 Linear regression plot of β vs. ε_{cf} relationship. (First published by Engineers Australia. Reprinted with permission.)

Accordingly, Equation (5.75) may be further simplified by considering two constant values of β for each range of ε_{cf} specified for α. The constants are selected to be the average of the two end values of each strain range (Rasheed and Motto 2010):

$$\beta = \frac{\beta_0 + \beta_{0.0015}}{2} = 0.3447 \qquad 0 \le \varepsilon_{cf} < 0.0015$$

$$\beta = \frac{\beta_{0.0015} + \beta_{0.003}}{2} = 0.3864 \qquad 0.0015 \le \varepsilon_{cf} < 0.003 \tag{5.76}$$

Substituting into the moment equilibrium Equation (5.62) for the first range of ε_{cf} values $0 \le \varepsilon_{cf} < 0.0015$, this leads to the following cubic equation:

$$A_3\left(\frac{c}{d_f}\right) + B_3\left(\frac{c}{d_f}\right)^2 + D_3\left(\frac{c}{d_f}\right)^3 + Q_2 = 0 \tag{5.77}$$

where if $\varepsilon_{cf} < 0.0015$,

$$A_3 = 0.3447Q_1(1-\Psi_f) - Q_2 - 0.0417\Psi_f \tag{5.78}$$

$$B_3 = 0.0561\Psi_f - 0.3447Q_1(1-\Psi_f) - 366.67\Psi_f\bar{\varepsilon}_{fu}^{\max} \tag{5.79}$$

$$D_3 = 126.3911\Psi_f\bar{\varepsilon}_{fu}^{\max} - 0.0144\Psi_f \tag{5.80}$$

Similarly, if $0.0015 \le \varepsilon_{cf} \le 0.003$, Equation (5.77) holds with:

$$A_3 = 0.3864Q_1(1-\Psi_f) - Q_2 - 0.4042\Psi_f \tag{5.81}$$

$$B_3 = 0.5604\Psi_f - 0.3864Q_1(1-\Psi_f) - 125\Psi_f\bar{\varepsilon}_{fu}^{\max} \tag{5.82}$$

$$D_3 = 48.3\Psi_f\bar{\varepsilon}_{fu}^{\max} - 0.1562\Psi_f \tag{5.83}$$

Once $\frac{c}{d_f}$ is determined, ρ_f can be calculated from the force equilibrium:

$$\rho_f = \alpha\frac{f_c'}{0.9f_{fu}}\frac{c}{d} - \rho_s\frac{f_y}{0.9f_{fu}} \tag{5.84}$$

5.4.3.4 Linear Regression Solution for Rupture Failure Mode

Alternatively, Rasheed and Motto (2010) derived a statistically accurate linear relationship between the strengthening ratio and the reinforcement force ratio based on a parametric study of singly and doubly-reinforced strengthened rectangular sections, as seen in Figure 5.14. This parametric study had 516 data points yielding an $R^2 = 0.9994$ which represents perfect linearity. However, Ψ_f was not considered in that equation. This linear equation may be equally used in analysis and design. The

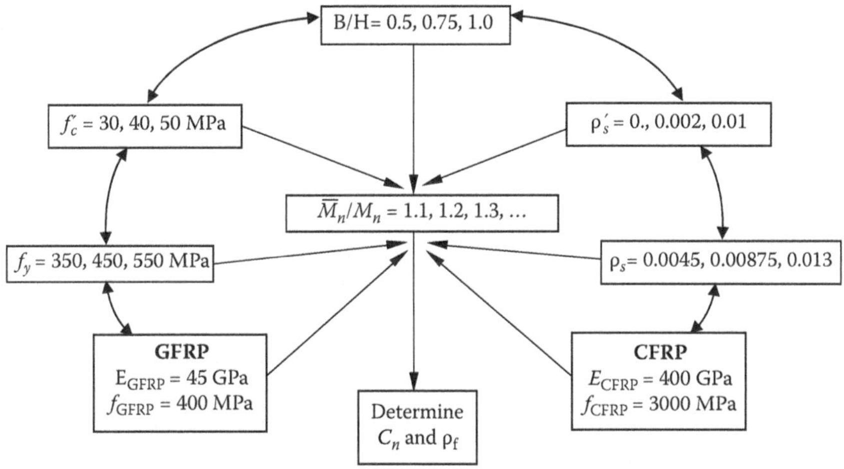

FIGURE 5.14 Variation of design variables in the parametric study of Rasheed and Motto (2010). (First published by Engineers Australia. Reprinted with permission.)

reinforcement force ratio is $\lambda = \frac{\rho_f \times 0.9 f_{fu}}{\rho_s f_y}$, while the strengthening ratio is $\frac{\bar{M}_n}{M_n} = \frac{\bar{M}_u}{M_u}$ if $\bar{\phi} = 0.9$:

$$\lambda = 0.9626 \frac{\bar{M}_n}{M_n} - 0.976 \tag{5.85}$$

In a follow-up study, Saqan, Rasheed, and Hawileh (2013) derived a similar statistically accurate linear relationship while considering $\Psi_f = 0.85$, as per ACI 440.2R-08. However, their definition of the $\rho_f = \frac{A_f}{bd_f}$ differs from the definition of the same variable $\rho_f = \frac{A_f}{bd}$ in this textbook. Accordingly, the same relationship, developed by Saqan, Rasheed, and Hawileh (2013), is rederived here using the latter definition of the FRP reinforcement ratio. This linear relationship correlates 177 data points of beam section designs performed in accordance to ACI 440.2R-08 and yielding an $R^2 = 0.9973$, as seen in Figure 5.15. The *x*-axis has the FRP effective force $\left(\frac{\rho_f f_f}{\rho_s f_y}\right) \times \frac{d_f}{d}$, while the *y*-axis has the strengthening ratio $\frac{\bar{M}_n}{M_n}$:

$$\frac{\bar{M}_n}{M_n} = 0.7815 \left(\frac{\rho_f f_f}{\rho_s f_y}\right) \times \frac{d_f}{d} + 1 \tag{5.86}$$

To prove the accuracy of this linear relationship, the formulation developed by Saqan, Rasheed, and Hawileh (2013) will be followed here.
Moment equilibrium:

$$\bar{M}_n = A_s f_y \left(d - \frac{\beta_1 c}{2}\right) + \psi_f A_f f_f \left(d_f - \frac{\beta_1 c}{2}\right) \tag{5.87}$$

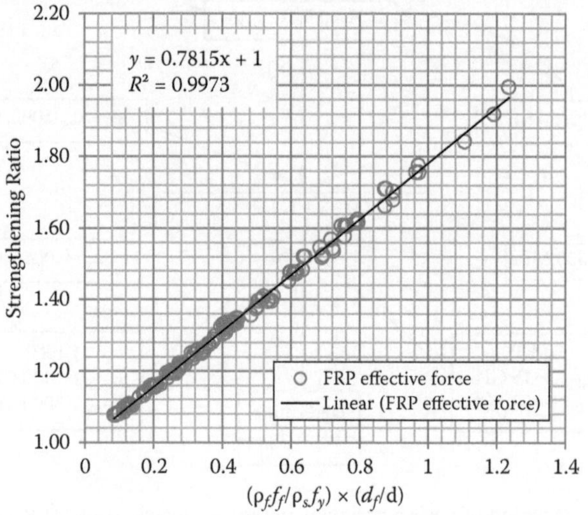

FIGURE 5.15 Linear correlation of design data derived after the work of Saqan, Rasheed, and Hawileh (2013).

$$M_n = A_s f_y \left(d - \frac{(\beta_1 c)_{un}}{2} \right) \tag{5.88}$$

Dividing $\bar{M}_n$ from Equation (5.87) by M_n from Equation (5.88),

$$\frac{\bar{M}_n}{M_n} = \frac{A_s f_y \left(d - \frac{\beta_1 c}{2} \right) + \psi_f A_f f_f \left(d_f - \frac{\beta_1 c}{2} \right)}{A_s f_y \left(d - \frac{(\beta_1 c)_{un}}{2} \right)} \tag{5.89}$$

Substituting the steel ratio $\rho_s = A_s/bd$ and the FRP ratio $\rho_f = A_f/bd$ in Equation (5.89) and rearranging, the following expression is derived:

$$\frac{\bar{M}_n}{M_n} = \frac{\left(1 - \frac{\beta_1 c}{2d} \right)}{\left(1 - \frac{(\beta_1 c)_{un}}{2d} \right)} + \psi_f \frac{\rho_f f_f}{\rho_s f_y} \left(\frac{d_f}{d} \right) \frac{\left(1 - \frac{\beta_1 c}{2d_f} \right)}{\left(1 - \frac{(\beta_1 c)_{un}}{2d} \right)} \tag{5.90}$$

In the limit when $\bar{M}_n$ approaches M_n, the first term of Equation (5.90) approaches 1. The quotient in the second term $\left(1 - \frac{\beta_1 c}{2d_f}\right) / \left(1 - \frac{(\beta_1 c)_{un}}{2d}\right)$ is found to have very little variation. Accordingly, Equation (5.90) closely represents a linear relationship. This relationship may be directly used in design irrespective of the failure mode admitted. Accordingly, it will be used to determine the FRP reinforcement ratio ρ_f directly by substituting $f_f = 0.9 f_{fu}$ in Equation (5.86).

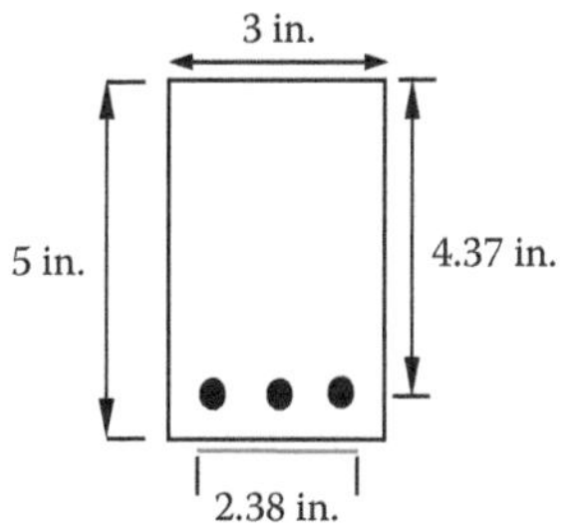

FIGURE 5.16 Example 5.10 showing the section and FRP design details.

Example 5.10: Analysis

A singly reinforced section was constructed and tested by Triantafillou and Plevris (1992). CFRP was used having E_f = 26,980 ksi and f_{fu}^* = 210 ksi. The beam designation was Beam 3 with the following section properties, as seen in Figure 5.16:

$$f_c' = 6.48 \text{ ksi} \qquad b_f = 2.38''$$

$$f_y = 75 \text{ ksi} \qquad t_f = 0.0079''$$

$$A_s = 0.051 \text{ in}^2 \qquad \varepsilon_{bi} = 0$$

Determine the design moment capacity of the section and compare it to the experimental moment to have a feel for the strength reduction factors warranted by ACI 440.

Solution:

Determine the failure mode first.

$$\beta_1 = 1.05 - 0.05\frac{6480}{1000} = 0.726$$

$$\bar{\varepsilon}_{fu}^{\max} = 0.9C_E\varepsilon_{fu}^* + \varepsilon_{bi} = 0.9 \times 0.95 \times \frac{210}{26980} + 0 = 0.00665$$

$$a_b^{\max} = \beta_1\frac{\varepsilon_{cu} \times d_f}{\varepsilon_{cu} + \bar{\varepsilon}_{fu}^{\max}} = 0.726 \times \frac{0.003 \times 5}{0.003 + 0.00665} = 1.13''$$

$$0.9f_{fu} = 0.9C_E f_{fu}^* = 0.9 \times 0.95 \times 210 = 179.55 \text{ ksi}$$

$$\rho_f^{b\max} = 0.85\frac{f_c'}{0.9f_{fu}}\frac{a_b^{\max}}{d} - \rho_s\frac{f_y}{0.9f_{fu}}$$

$$\rho_f^{b\max} = 0.85 \times \frac{6.48}{179.55} \times \frac{1.13}{4.37} - \frac{0.051}{3 \times 4.37} \times \frac{75}{179.55} = 0.00631$$

$$\rho_f = \frac{2.38 \times 0.0079}{3 \times 4.37} = 0.00143 < \rho_f^{b\max} = 0.00631$$

The failure mode seems to be FRP rupture:

$$\alpha f_c' bc = A_s f_y + A_f(0.9 f_{fu})$$

$$\varepsilon_{cf} = \frac{\bar{\varepsilon}_{fu}^{max} c}{d_f - c} = 0.00665\frac{c}{5-c}$$

$$\varepsilon_c' = 1.71\frac{f_c'}{E_c} = 3\times10^{-5}\sqrt{6480} = 0.00241$$

$$\alpha = \frac{\varepsilon_{cf}}{\varepsilon_c'} - \frac{1}{3}\left(\frac{\varepsilon_{cf}}{\varepsilon_c'}\right)^2 = 2.76\frac{c}{5-c} - \frac{1}{3}\left(2.76\frac{c}{5-c}\right)^2$$

$$\alpha = 2.76\frac{c}{5-c} - 2.54\frac{c^2}{(5-c)^2}$$

$$\alpha \times 6.48 \times 3c = 0.051\times 75 + 2.38\times 0.0079\times 179.55 = 7.2$$

$$\left[2.76\frac{c}{(5-c)} - 2.54\frac{c^2}{(5-c)^2}\right]\times 19.44c = 7.2 \qquad \text{multiply by } (5-c)^2$$

$$53.65c^2(5-c) - 49.38c^3 = 7.2(5-c)^2$$

$$103.03c^3 - 268.25c^2 + 180 - 72c + 7.2c^2 = 0$$

$$103.03c^3 - 261.05c^2 - 72c + 180 = 0$$

$$c = 0.828'' \quad \text{(Goal-Seek in Excel)}$$

$$\varepsilon_{cf} = 0.00665\frac{0.828}{5-0.828} = 0.00132 < 0.003$$

$$\varepsilon_s = 0.00665\frac{4.37-0.828}{5-0.828} = 0.00565 > 0.005 \Rightarrow \bar{\phi} = 0.9$$

$$\varepsilon_{fd} = 0.083\sqrt{\frac{6480}{1\times 26980000\times 0.0079}} = 0.0145$$

$$\bar{\varepsilon}_{fu}^{max} = 0.00665 < \varepsilon_{fd}$$

Thus, FRP rupture is confirmed

$$\bar{M}_u = \bar{\phi} A_s f_y\left(d - \frac{\beta_1 c}{2}\right) + \bar{\phi}\Psi_f A_f \bar{f}_{fu}^{max}\left(d_f - \frac{\beta_1 c}{2}\right)$$

$$\bar{f}_{fu}^{max} = E_f \times \bar{\varepsilon}_{fu}^{max} = 26980\times 0.00665 = 179.42 \approx 0.9 f_{fu}$$

$$\bar{M}_u = 0.9\times 0.051\times 75\times\left(4.37 - \frac{0.726\times 0.828}{2}\right)$$

$$+ 0.9\times 0.85\times 2.38\times 0.0079\times 179.42\times\left(5 - \frac{0.726\times 0.828}{2}\right)$$

$$= 14.01 + 12.13 = 26.14 \text{ k-in} = 2.178 \text{ k-ft}$$

from the paper, $\bar{M}_n = 34.94$ k-in $= 2.912$ k-ft

Thus, the average actual strength reduction factor is $\frac{\bar{M}_u}{\bar{M}_{nexp}} = \frac{26.14}{34.94} = 0.748 < 0.9.$

What are the sources of strength reduction?
These are

$1 - \bar{\phi}$
$2 - \Psi_f$
$3 - 0.9f_{fu}$
$4 - C_E$
$5 - f_{fu} = f_{fu}^* - 3\sigma$

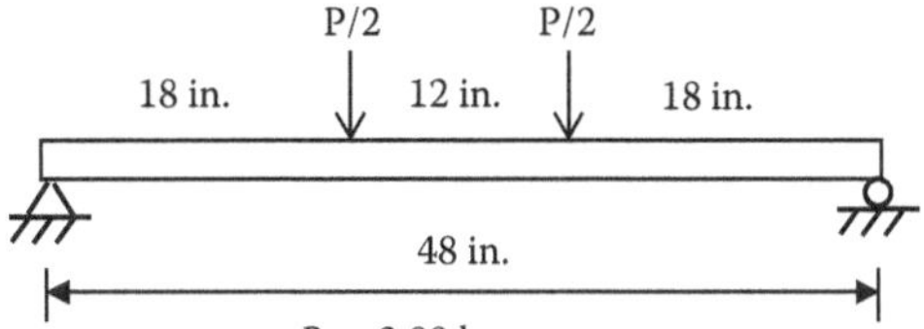

$P_n = 3.88$ k

$\bar{M}_n \frac{P_n}{2} \times a = 34.94$ k - in

Now, let's calculate M_u without strengthening:

$$a = \frac{A_s f_y}{0.85 f_c' b} = \frac{0.051 \times 75}{0.85 \times 6.48 \times 3} = 0.231''$$

$$M_u = \phi A_s f_y \left(d - \frac{a}{2} \right) = 0.9 \times 0.051 \times 75 \left(4.37 - \frac{0.231}{2} \right) = 14.65 \text{ k-in} = 1.22 \text{ k-ft}$$

Thus, the moment strengthening ratio is

$$\frac{\bar{M}_u}{M_u} = \frac{26.14}{14.65} = 1.78$$

Example 5.11: Design

A doubly reinforced section was built and tested by Arduini, Tommaso, and Nanni (1997), illustrated in Figure 5.17. The beam was designated as Beam B2, which failed by rupture of FRP. In this example, we shall design for FRP to achieve $\bar{M}_n$ from experiment. Then we will compare the actual FRP area to what we obtain.

$\bar{M}_{nexp} = 93.5$ kN-m ($\bar{P}_n = 170$ kN) $\quad E_f = 400$ GPa
$M_{nexp} = 51.66$ kN-m ($P_n = 93.93$ kN) $\quad f_{fu}^* = 3000$ MPa
$A_s = 398\ \text{mm}^2 \quad A_s' = 265\ \text{mm}^2$
$f_y = 340$ MPa $\quad f_c' = 30$ MPa
actual $b_f = 300$ mm $\quad t_f = 0.17$ mm
Assume 1 FRP layer only.

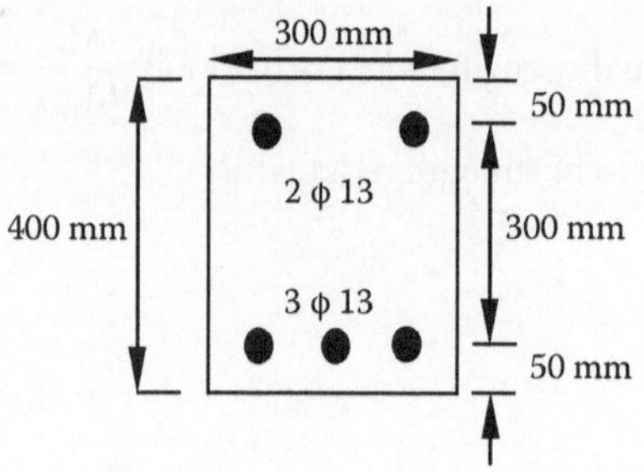

FIGURE 5.17 Example 5.11 showing the beam section tested by Arduini, Tommaso, and Nanni (1997).

Solution:

The failure mode is unknown a priori. Assume ductile concrete crushing (tension controlled). Ignore the compression steel for now:

$$d_t = 350 \text{ mm} = d$$

$$d' = 50 \text{ mm}$$

$$d_f = 400.09 \text{ mm} = h + \frac{t_f}{2}$$

$$\rho_s = \frac{\frac{\pi}{4} \times 13^2 \times 3}{300 \times 350} = 0.00379$$

$$Q_1 = \rho_s \frac{f_y}{f'_c} \frac{d}{d_f} = 0.00379 \times \frac{340}{30} \times \frac{350}{400.09} = 0.0376$$

$$Q_2 = \frac{\bar{M}_u}{\phi f'_c b d_f^2} + Q_1\left(\Psi_f - \frac{d}{d_f}\right) = \frac{93.5 \times 10^3 \times 10^3}{30 \times 300 \times 400.09^2}$$

$$+ 0.0376\left(0.85 - \frac{350}{400.09}\right) = 0.06397$$

$$\left(\frac{a}{d_f}\right)^2 - (2 - 0.208 \times 0.0376)\left(\frac{a}{d_f}\right) + 2.77 \times 0.06397 = 0$$

$$\left(\frac{a}{d_f}\right)^2 - 1.992\left(\frac{a}{d_f}\right) + 0.1772 = 0$$

$$\frac{a}{d_f} = \frac{1.992 - \sqrt{1.992^2 - 4 \times 0.1772}}{2} = 0.0933$$

$$a = 37.34 \text{ mm} \qquad \beta_1 = 0.85$$

$$c = \frac{a}{\beta_1} = 43.93 \text{ mm}$$

$$\bar{\varepsilon}_f = \varepsilon_{fe} = \frac{0.003 \times (400.09 - 43.93)}{43.93} = 0.0243 > \varepsilon_{fu} = \frac{3000}{400000} = 0.0075$$

Thus, the failure mode is expected to be rupture of FRP.
Assuming interior exposure for CFRP,

$$0.9f_{fu} = 0.9C_E f_{fu}^* = 0.9 \times 0.95 \times 3000 = 2565 \text{ MPa}$$

Finding ε_{bi} based on the self-weight of the beam only (experimentally tested beam),

$$w_{DL} = 23.5 \text{ kN/m}^3 \times \frac{300 \times 400}{10^6} = 2.82 \text{ kN/m}$$

$$\text{M}_{\text{DL}} = \frac{w_{DL} \times L_n^2}{8} = \frac{2.82 \times 2.5^2}{8} = 2.203 \text{ kN-m} < \text{M}_{\text{cr}}$$

$$M_{cr} = \frac{f_r I_{gt}}{y_{bot}} = \frac{0.62\sqrt{30} \times 1.7 \times 10^9}{(400 - 201.1)} = 29024614 \text{ N-mm} = 29 \text{ kN-m}$$

$$y_{top} = \frac{300 \times 400 \times 200 + (n-1) \times 265.5 \times 50 + (n-1) \times 398.2 \times 350}{300 \times 400 + (n-1) \times 265.5 + (n-1) \times 398.2} = 201.1 \text{ mm}$$

$$n = \frac{200000}{4700\sqrt{30}} = 7.77$$

$$I_{gt} = \frac{300 \times 400^3}{12} + 300 \times 400 \times (200 - 201.1)^2 + (n-1) \times 265.5 \times (201.1 - 50)^2$$

$$+ (n-1) \times 398.2 \times (201.1 - 350)^2 = 1.7 \times 10^9 \text{ mm}^4$$

$$\varepsilon_{bi} = \frac{M_{DL} \times (h - y_{top})}{E_c I_{gt}} = \frac{2.203 \times 10^6 \times (400 - 201.1)}{4700\sqrt{30} \times 1.7 \times 10^9} = 0.00001 \rightarrow \text{ negligible}$$

Assuming the beam loaded with dead load only during strengthening,

$$\bar{\varepsilon}_{fu}^{\max} = 0.9C_E \varepsilon_{fu}^* + \varepsilon_{bi} = 0.9 \times 0.95 \times \frac{3000}{400000} + 0.00001 = 0.00642$$

$$\varepsilon_c' = 1.71\frac{f_c'}{E_c} = 1.71 \times \frac{30}{4700\sqrt{30}} = 0.00199 \approx 0.002$$

First, use the exact fifth-degree polynomial ignoring compression steel:

$$Q_1 = \rho_s \frac{f_y}{f_c'} \frac{d}{d_f} = 0.0376$$

$$Q_2 = \frac{\bar{M}_u}{\phi f_c' b d_f^2} + Q_1\left(\Psi_f - \frac{d}{d_f}\right) = 0.06397$$

$$A = 3(Q_1(1 - \Psi_f) - 9Q_2)\varepsilon_c'^3 - 3Q_2\varepsilon_c'^2 \bar{\varepsilon}_{fu}^{\max}$$

$$= 3(0.0376 \times 0.15 - 9 \times 0.06397) \times 0.002^3 - 3 \times 0.06397 \times 0.002^2 \times 0.00642$$

$$= -1.861 \times 10^{-8}$$

$$B = 9\left(3Q_2 - Q_1(1-\Psi_f)\right)\varepsilon_c'^3 + 3\left(2Q_2 - \frac{Q_1}{4}(1-\Psi_f) - 3\Psi_f\right)\varepsilon_c'^2\bar{\varepsilon}_{fu}^{max}$$

$$= 9(3\times 0.06397 - 0.0376\times 0.15)\times 0.002^3$$

$$+3\left(2\times 0.06397 - \frac{0.0376}{4}\times 0.15 - 3\times 0.85\right)\times 0.002^2\times 0.00642 = -1.733\times 10^{-7}$$

$$D = 9\left(Q_1(1-\Psi_f) - Q_2\right)\varepsilon_c'^3 + 3\left(\frac{Q_1}{2}\left(1-\frac{\Psi_f}{2}\right) + 6\Psi_f - Q_2\right)\varepsilon_c'^2\bar{\varepsilon}_{fu}^{max} + 6\Psi_f\varepsilon_c'\bar{\varepsilon}_{fu}^{max\,2}$$

$$= 9(0.0376\times 0.15 - 0.06397)\times 0.002^3$$

$$+3\left(\frac{0.0376}{2}\times 0.575 + 6\times 0.85 - 0.06397\right)\times 0.002^2\times 0.00642$$

$$+6\times 0.85\times 0.002\times 0.00642^2$$

$$= 8.05\times 10^{-7}$$

$$E = -3Q_1(1-\Psi_f)\varepsilon_c'^3 - 3\left(\frac{Q_1}{4}(1-\Psi_f) + 5\Psi_f\right)\varepsilon_c'^2\bar{\varepsilon}_{fu}^{max} - \frac{31}{4}\Psi_f\varepsilon_c'\bar{\varepsilon}_{fu}^{max\,2} - \Psi_f\bar{\varepsilon}_{fu}^{max\,3}$$

$$= -3\times 0.0376\times 0.15\times 0.002^3 - 3\left(\frac{0.0376}{4}\times 0.15 + 5\times 0.85\right)\times 0.002^2\times 0.00642$$

$$-\frac{31}{4}\times 0.85\times 0.002\times 0.00642^2 - 0.85\times 0.00642^3 = -1.0956\times 10^{-6}$$

$$F = 3\Psi_f\varepsilon_c'^2\bar{\varepsilon}_{fu}^{max} + \frac{7}{4}\Psi_f\varepsilon_c'\bar{\varepsilon}_{fu}^{max\,2} + \frac{1}{4}\Psi_f\bar{\varepsilon}_{fu}^{max\,3}$$

$$= 3\times 0.85\times 0.002^2\times 0.00642 + \frac{7}{4}\times 0.85\times 0.002\times 0.00642^2 + \frac{1}{4}\times 0.85\times 0.00642^3$$

$$= 2.4433\times 10^{-7}$$

$$9Q_2\varepsilon_c'^3 = 9\times 0.06397\times 0.002^3 = 4.60584\times 10^{-9}$$

The fifth-degree polynomial equation becomes

$$4.60584\times 10^{-9} - 1.861\times 10^{-8}\frac{c}{d_f} - 1.733\times 10^{-7}\left(\frac{c}{d_f}\right)^2 + 8.05\times 10^{-7}\left(\frac{c}{d_f}\right)^3$$

$$-1.0956\times 10^{-6}\left(\frac{c}{d_f}\right)^4 + 2.4433\times 10^{-7}\left(\frac{c}{d_f}\right)^5 = 0$$

Multiplying the previous equation by 1×10^9,

$$4.60584 - 18.61\frac{c}{d_f} - 173.3\left(\frac{c}{d_f}\right)^2 + 805\left(\frac{c}{d_f}\right)^3 - 1095.6\left(\frac{c}{d_f}\right)^4 + 244.33\left(\frac{c}{d_f}\right)^5 = 0$$

$$\frac{c}{d_f} = 0.1532 \Rightarrow c = 61.29\text{ mm} \Rightarrow \varepsilon_{cf} = \frac{c}{d_f - c}\bar{\varepsilon}_{fu}^{max} = 0.001161$$

Applying the approximate solution equation,

For $\varepsilon_{cf} < 0.0015$

$$A_3 = 0.3447 \times 0.0376 \times 0.15 - 0.06397 - 0.0417 \times 0.85 = -0.0975$$

$$B_3 = 0.0561 \times 0.85 - 0.3447 \times 0.0376 \times 0.15 - 366.67 \times 0.85 \times 0.00642 = -1.9552$$

$$D_3 = 126.3911 \times 0.85 \times 0.00642 - 0.0144 \times 0.85 = 0.67748$$

$$0.06397 - 0.0975\left(\frac{c}{d_f}\right) - 1.9552\left(\frac{c}{d_f}\right)^2 + 0.67748\left(\frac{c}{d_f}\right)^3 = 0$$

$$\frac{c}{d_f} = 0.1607$$

$$c = 64.29 \text{ mm}$$

$$\rho_f = \alpha \frac{f_c'}{0.9 f_{fu}} \frac{c}{d} - \rho_s \frac{f_y}{0.9 f_{fu}}$$

$$\varepsilon_{cf} = \frac{c}{d_f - c} \bar{\varepsilon}_{fu}^{max} = \frac{64.29}{400.09 - 64.29} \times 0.00642 = 0.00123 \quad (\text{exact} = 0.00116)$$

$$\alpha = \frac{\varepsilon_{cf}}{\varepsilon_c'} - \frac{\varepsilon_{cf}^2}{3\varepsilon_c'^2} = 0.4889 \quad (\text{exact } 0.4679)$$

$$\alpha_{approx} = 500 \times \varepsilon_{cf} - 83333 \times \varepsilon_{cf}^2 = 0.4889$$

$$\rho_f = 0.4889 \times \frac{30}{2565} \times \frac{64.29}{350} - 0.003792 \times \frac{340}{2565} = 5.477 \times 10^{-4} \quad (\text{exact} = 4.557 \times 10^{-4})$$

$$A_f = \rho_f bd = 57.51 \text{ mm}^2 \qquad (\text{exact} = 47.85 \text{mm}^2)$$

$$t_f = \frac{A_f}{b_f} = \frac{57.51}{300} = 0.192 \text{ mm} > t_{f\,actual} = 0.17 \text{ mm O.K.}$$

(exact = 0.16 mm) for singly reinforced section (with doubly $t_f > 0.17$ is expected)

The third approach is by using the statistically correlated linear equation:
Unstrengthened beam:

$$0.85 f_c' b \beta_1 c + A_s' E_s \frac{0.003}{c}(c - d') = A_s f_y$$

$$6502.5c^2 + 265.5 \times 600c - 265.5 \times 600 \times 50 - 398.2 \times 340c = 0$$

$$6502.2c^2 + 23912c - 7965000 = 0$$

$$c = \frac{-23912 + \sqrt{23912^2 + 4 \times 6502.5 \times 7965000}}{2 \times 6502.5} = 33.21 \text{ mm}$$

$$f_s' = E_s \times \frac{0.003}{33.21}(33.21 - 50) = -303.34 \text{ MPa}$$

$$M_n = A_s f_y\left(d - \frac{\beta_1 c}{2}\right) + A_s' f_s'\left(\frac{\beta_1 c}{2} - d'\right)$$

$$M_n = 398.2 \times 340 \times \left(350 - \frac{0.85 \times 33.21}{2}\right) + 265.5 \times (-303.34) \times \left(\frac{0.85 \times 33.21}{2} - 50\right)$$

$$= 48365044.86 \text{ N-mm} = 48.365 \text{ kN-m}$$

$$\frac{\bar{M}_n}{M_n} = \frac{\bar{M}_u}{M_u} = \frac{93.5}{48.365} = 1.9332$$

$$\frac{\bar{M}_n}{M_n} = 0.7815 \frac{\rho_f f_f}{\rho_s f_y} \frac{d_f}{d} + 1 = 1.9332$$

$$\lambda = \frac{\rho_f \times 0.9 f_{fu}}{\rho_s f_y} \frac{d_f}{d} = \frac{0.9332}{0.7815} \Rightarrow \rho_f = \frac{1.194 \times 0.003792 \times 340 \times 350}{2565 \times 400.09} = 5.25 \times 10^{-4}$$

$$A_f = \rho_f bd = 55.13 \text{ mm}^2$$

$$t_f = \frac{A_f}{b_f} = \frac{55.13}{300} = 0.184 \text{ mm} > t_{factual} = 0.17 \text{ mm}$$

Note: The t_f is slightly greater than 0.17 mm, since the equation is correlated with Ψ_f.

Check the debonding strain:

$$\varepsilon_{fd} = 0.41\sqrt{\frac{30}{1 \times 400000 \times 0.184}} = 0.008278 > \bar{\varepsilon}_{fu}^{max} = 0.00642$$

Thus, the failure mode of rupture of FRP is confirmed.

Example 5.12: Design of One-Way Solid Slab

Design the slab given in Figure 5.18 based on ACI 318-83 code, and then strengthen it based on ACI 440.2R-08, assuming no change in dead load.

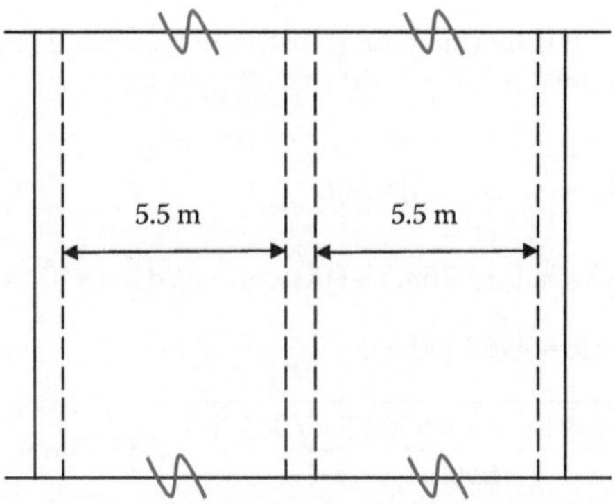

FIGURE 5.18 Example 5.12 showing the top view of the solid slab.

The slab is a two-span, continuous, solid, one-way slab with the following parameters:

$w_{LL} = 4.8$ kN/m^2
$w_{LLupgrade} = 7.0$ kN/m^2
$f_y = 400$ MPa
$f'_c = 25$ MPa

Solution:

Let $h_{min} = \frac{L_n}{28} = \frac{5.5}{28} = 0.196$ m
Use $h = 0.2$ m = 200 mm
DL self wt: $0.2 \times 23.6 = 4.72$ kN/m/m
DL tiling: $0.05 \times 17 = 0.85$ kN/m/m
DL total: $W_{DL} = 5.57$ kN/m^2
LL total: $W_{LL} = 4.8$ kN/m^2

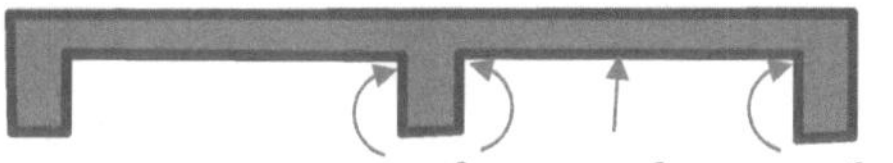

As per ACI 318-83:

$$w_u = 1.4w_{DL} + 1.7w_{LL}$$

$$= 1.4 \times 5.57 + 1.7 \times 4.8 = 15.96 \text{ kN/m}^2$$

$d = 200 - 25 = 175$ mm

$$\text{M\_@interior\_support} = \frac{w_u L_n^2}{9} = 53.64 \text{ kN-m/m}$$

$$R = \frac{M_u}{\phi f_c' b d^2} = \frac{53.64 \times 10^6 N - mm/m}{0.9 \times 25 \times 1000 \times 175^2} = 0.0778$$

$$\omega = \frac{1 - \sqrt{1 - 2.36R}}{1.18} = 0.0818$$

$$\rho_s = \omega \frac{f_c'}{f_y} = 0.0818 \times \frac{25}{400} = 0.00511$$

$$A_{s1} = \rho_s bd = 0.00511 \times 1000 \times 175 = 895 \text{ mm}^2/\text{m}$$

Try No. 13 bars (Area = 133 mm^2).

$$\text{No. of bars/m} = \frac{895}{133} = 6.73$$

$$\text{Spacing} = \frac{1000}{6.73} = 148.6 \text{ mm} \approx 150 \text{ mm} \Rightarrow \text{use No. 13 bars @ 150 mm}$$

$$M_+ \text{@midspan} = \frac{w_u L_n^2}{14} = 34.49 \text{ kN-m/m}$$

$$R = \frac{34.49 \times 10^6 N - mm/m}{0.9 \times 25 \times 1000 \times 175^2} = 0.050$$

$$\omega = 0.0516$$

$$\rho_s = 0.00323$$

$$A_{s2} = 565 \text{ mm}^2/\text{m}$$

Try No. 10 bars (Area = 100 mm²).

$$\text{No. of bars/m} = \frac{565}{100} = 5.65$$

$$\text{Spacing} = \frac{1000}{5.65} = 177 \text{ mm} \quad \text{use No. 10 bars @ 150 mm}$$

$$\text{M\_exterior\_support} = \frac{w_u L_n^2}{24} = 20.12 \text{ kN-m/m}$$

$$R = 0.0292$$

$$\omega = 0.0297$$

$$\rho_s = 0.00186$$

$$A_{s3} = 325 \text{ mm}^2/\text{m}$$

$$\text{Check } \rho_{max} = 0.75 \times 0.85\beta_1 \frac{f_c'}{f_y} \frac{600}{600 + f_y}$$

$$\beta_1 = 0.85 \text{ since } f_c' < 30 \text{ MPa}$$

$$\rho_{max} = 0.0203 > \text{ all } \rho_s$$

$$A_{s\min} = \rho_{\min} bh = 0.0018 \times 1000 \times 200 = 360 \text{ mm}^2/\text{m}$$

$$A_{s3} < A_{s\min} \text{ use } A_{s3} = A_{s\min} = 360 \text{ mm}^2/\text{m}$$

Try No. 10 bars (Area = 100 mm²).

No. of bars/m = 3.6

$$\text{Spacing} = \frac{1000}{3.6} = 278 \text{ mm} \Rightarrow \text{use No. 10 bars @ 250 mm}$$

$$s_{max} \leq \min(3h, 450 \text{ mm})$$

$$s_{max} = 450 \text{ mm}$$

In the transverse direction, use temperature and shrinkage reinforcement:

$A_{s\min} = 360\ \text{mm}^2/\text{m}$

Use No. 10 bars @ 250 mm $(400\ \text{mm}^2/\text{m})$.

Check shear capacity.

$$V_u = 1.15\frac{w_u L_n}{2} = 50.47\ \text{kN}$$

$$V_{ud} = V_u - w_u d = 50.47 - 15.96 \times 0.175 = 47.68\ \text{kN}$$

$$\frac{V_{ud}}{\phi} = \frac{47.68}{0.85} = 56.1\ \text{kN}$$

$$V_c = 0.17\sqrt{f_c'}b_w d = 0.17\sqrt{25} \times 1000 \times 175 = 148{,}750\ \text{N} = 148.75\ \text{kN} >> \frac{V_{ud}}{\phi}\ \text{O.K.}$$

Strengthening design:

$$\bar{w}_u = 1.2w_{DL} + 1.6w_{LL} = 1.2 \times 5.57 + 1.6 \times 7.0 = 17.884\ \text{kN/m}^2$$

There is an advantage of load factors.
Strengthening limits:

$$w_{SL} = 1.1 \times w_{DL} + 0.75 \times w_{LLnew} = 1.1 \times 5.57 + 0.75 \times 7.0$$

$$= 11.38\ \text{kN/m}^2 < 15.96\ \text{kN/m}^2\ \ \text{O.K.}$$

since $w_{LLnew} = 7\ \text{kN/m}^2 = 146\ \text{psf} < 150\ \text{psf}$

Positive moment section:

No. 10 @ 150 mm
No. of bars/m = $\frac{1000}{150} = 6.67$ bars/m
Area of reinforcement/m = 100 $\text{mm}^2 \times 6.67 = 667\ \text{mm}^2/\text{m}$

$$\rho_s = \frac{A_{s+}}{bd} = \frac{667}{1000 \times 175} = 0.00381$$

$$\omega = \rho_s \frac{f_y}{f_c'} = 0.00381 \times \frac{400}{25} = 0.06095$$

$$R = \omega - 0.59\omega^2 = 0.05876 = \frac{M_n}{f_c' bd^2} \Rightarrow$$

$$M_n = 0.05876 \times 25 \times 1000 \times 175^2 = 44988444.5\ \text{N-mm/m}$$

$$M_n = 44.99\ \text{kN-m/m}$$

$$\bar{M}_u = \frac{\bar{w}_u L_n^2}{14} = \frac{17.884 \times 5.5^2}{14} = 38.64\ \text{kN-m/m}$$

$$\phi M_n = 0.9 \times 44.99 = 40.5\ \text{kN-m/m} > \bar{M}_u = 38.64\ \text{kN-m/m}$$

No need to strengthen the positive moment region.
Negative moment section at interior support:

No. 13 @ 150 mm
No. of bars/m = $\frac{1000}{150} = 6.67$ bars/m
Area of reinforcement/m = 133 mm² × 6.67 = 887 mm²/m

$$\rho_s = \frac{A_{s-}}{bd} = \frac{887}{1000 \times 175} = 0.00507$$

$$\omega = \rho_s \frac{f_y}{f_c'} = 0.00507 \times \frac{400}{25} = 0.0811$$

$$R = \omega - 0.59\omega^2 = 0.07723 = \frac{M_n}{f_c' b d^2}$$

$$M_n = 0.07723 \times 25 \times 1000 \times 175^2 = 59{,}126{,}123.4 \text{ N-mm/m} = 59.13 \text{ kN-m/m}$$

$$\bar{M}_u = \frac{\bar{w}_u L_n^2}{9} = \frac{17.884 \times 5.5^2}{9} = 60.11 \text{ kN-m/m}$$

$$\phi M_n = 0.9 \times 59.13 = 53.22 \text{ kN-m/m} < \bar{M}_u = 60.11 \text{ kN-m/m}$$

Thus, strengthening is needed at the interior negative region:

Use Aslan 500 NSM CFRP Tape.

Dimensions: 16 mm x 2 mm

$f_{fu} = 2068$ MPa $\qquad E_f = 124$ GPa $\qquad \varepsilon_{fu} = 0.0167$

Assume *only DL* at the time of strengthening:

$$M_{DL} = \frac{w_{DL} L_n^2}{9} = \frac{5.57 \times 5.5^2}{9} = 18.72 \text{ kN-m/m}$$

$$M_{cr} = \frac{f_r I_{gt}}{y_{top}}$$

Assuming a singly reinforced section:

$$y_{top} = \frac{1000 \times 200 \times 100 + (8.51 - 1) \times 887 \times 25}{1000 \times 200 + (8.51 - 1) \times 887} = 97.6 \text{ mm}$$

$$I_{gt} = \frac{1000 \times 200^3}{12} + 1000 \times 200 \times (100 - 97.6)^2 + (8.51 - 1) \times 887 \times (25 - 97.6)^2$$

$$= 702929149.2 \text{ mm}^4$$

$$= 7.02 \times 10^8 \text{ mm}^4$$

$$M_{cr} = \frac{0.62\sqrt{25} \times 7.02 \times 10^8}{97.6} = 22326643.1 \text{ N-mm/m} = 22.33 \text{ kN-m/m} > M_{DL}$$

Still the section is cracked under service load in its prior history.

$$k = \sqrt{2n\rho + \overline{n\rho}^2} - n\rho = 0.254$$

$$I_{cr} = \frac{b(kd)^3}{3} + nA_s(d - kd)^2 = \frac{1000 \times 44.41}{3} + 8.51 \times 887 \times (175 - 44.41)^2$$

$$= 157923864.6 = 1.579 \times 10^8 \text{ mm}^4$$

$$\varepsilon_{bi} = \frac{18.72 \times 10^6 (200 - 44.41)}{4700\sqrt{25} \times 4.1 \times 10^8} = 0.000785$$

$$\bar{\varepsilon}_{fd} = 0.7C_E\varepsilon_{fu} + \varepsilon_{bi} = 0.7 \times 0.95 \times 0.0167 + 0.000785 = 0.01189$$

Applying the linear regression equation:

$$\frac{\bar{M}_u}{M_u} = \frac{\bar{M}_n}{M_n} = \frac{60.11}{53.22} = 1.1295$$

$$\frac{\bar{M}_n}{M_n} = 0.7815 \frac{\rho_f f_f}{\rho_s f_y} \frac{d_f}{d} + 1 = 1.1295$$

$$\lambda = \frac{\rho_f \times 0.7 f_{fu}}{\rho_s f_y} \frac{d_f}{d} \Rightarrow \rho_f = \frac{0.1295 \times 0.00507 \times 400 \times 175}{0.7815 \times 0.7 \times 0.95 \times 2068 \times 200} = 0.000214$$

$$A_f = \rho_f bd = 37.41 \text{ mm}^2/\text{m}$$

Each NSM tape has the area of 32 mm^2.

$$\text{No. of NSM tapes/m} = \frac{37.41}{32} = 1.17 \text{ tapes}$$

$$\text{Spacing} = \frac{1000}{1.17} = 855 \text{ mm}$$

Use one NSM tape @ 850 mm.

$$\text{No. of tapes per meter} = \frac{1000}{850} = 1.176 \text{ tapes}$$

$$A_f = 32 \text{ mm}^2 \times 1.176 = 37.63 \text{ mm}^2/\text{m}$$

Using force equilibrium to find the neutral axis depth (c):

$$\alpha f_c' bc = A_s f_y + A_f \times 0.7 f_{fu}$$

$$\alpha = \frac{\varepsilon_{cf}}{\varepsilon_c'} - \frac{\varepsilon_{cf}^2}{3\varepsilon_c'^2} = \frac{\bar{\varepsilon}_{fd} \times c}{(d_f - c) \times \varepsilon_c'} - \frac{\bar{\varepsilon}_{fd}^2 \times c^2}{3(d_f - c)^2 \varepsilon_c'^2}$$

Into force equilibrium:

$$\left(\frac{\bar{\varepsilon}_{fd} \times c}{(d_f - c) \times \varepsilon_c'} - \frac{\bar{\varepsilon}_{fd}^2 \times c^2}{3(d_f - c)^2 \varepsilon_c'^2} \right) f_c' bc = A_s f_y + A_f \times 0.7 f_{fu}$$

Multiply this force equilibrium equation by $(d_f - c)^2 \varepsilon_c'^2$

$$\bar{\varepsilon}_{fd}\varepsilon_c'(d_f - c) f_c' bc^2 - \frac{\bar{\varepsilon}_{fd}^2}{3} f_c' bc^3 = (A_s f_y + A_f \times 0.7 f_{fu})(d_f - c)^2 \varepsilon_c'^2$$

$$\varepsilon_c' = 1.71 \times \frac{25}{4700\sqrt{25}} = 0.00182$$

$$0.01189 \times 0.00182 \times (200-c) \times 25 \times 1000 \times c^2 - \frac{0.01189^2}{3} \times 25 \times 1000c^3$$

$$= (887 \times 400 + 37.63 \times 0.7 \times 0.95 \times 2068) \times (200-c)^2 \times 0.00182^2$$

$$108.2c^2 - 0.541c^3 - 1.178c^3 - 1.347\left(40000 - 400c + c^2\right) = 0$$

$$-1.7191c^3 + 106.853c^2 + 538.8c - 53880 = 0$$

$$c = 25.18 \text{ mm}$$

$$\varepsilon_{cf} = \frac{\bar{\varepsilon}_{fd} \times c}{(d_f - c)} = \frac{0.01189 \times 25.18}{200 - 25.18} = 0.001713 < 0.003$$

So, debonding or limiting strain of NSM tape controls.

Another way to perform the design:

The approximate solution for $\varepsilon_{cf} > 0.0015$ is

$$A_3\left(\frac{c}{d_f}\right) + B_3\left(\frac{c}{d_f}\right)^2 + D_3\left(\frac{c}{d_f}\right)^3 + Q_2 = 0$$

$$Q_1 = \rho_s \frac{f_y}{f_c'} \frac{d}{d_f} = 0.00507 \times \frac{400}{25} \times \frac{175}{200} = 0.07098$$

$$Q_2 = \frac{\bar{M}_u}{\bar{\phi} f_c' b d_f^2} + Q_1\left(\Psi_f - \frac{d}{d_f}\right) = \frac{60.11 \times 10^6}{0.9 \times 25 \times 1000 \times 200^2}$$

$$+ 0.07098 \times \left(0.85 - \frac{175}{200}\right) = 0.065$$

$$A_3 = 0.3864 \times 0.07098 \times (1 - 0.85) - 0.065 - 0.4042 \times 0.85 = -0.4045$$

$$B_3 = 0.5604 \times 0.85 - 0.3864 \times 0.07098 \times (1 - 0.85) - 125 \times 0.85 \times 0.01189 = -0.7911$$

$$D_3 = 48.3 \times 0.85 \times 0.01189 - 0.1562 = 0.3319$$

$$\text{so,} \quad -0.4045\left(\frac{c}{d_f}\right) - 0.7911\left(\frac{c}{d_f}\right)^2 + 0.3319\left(\frac{c}{d_f}\right)^3 + 0.065 = 0$$

$$\left(\frac{c}{d_f}\right) = 0.1297 \Rightarrow c = 25.94 \text{ mm} \approx 25.18 \text{ mm}$$

$$\varepsilon_{cf} = \frac{\bar{\varepsilon}_{fd}}{d_f - c} c = \frac{0.01189}{200 - 25.94} \times 25.94 = 0.00177$$

$$\alpha = \frac{\varepsilon_{cf}}{\varepsilon_c'} - \frac{\varepsilon_{cf}^2}{3\varepsilon_c'^2} = \frac{0.00177}{0.00182} - \frac{0.001177^2}{3 \times 0.00182^2} = 0.657$$

or use $\alpha = 125 \times 0.00177 + 0.4042 = 0.625 \rightarrow$ more compatible

$$\rho_f = \alpha \frac{f_c'}{0.7 f_{fu}} \frac{c}{d} - \rho_s \frac{f_y}{0.7 f_{fu}} = 0.625 \frac{25}{0.7 \times 0.95 \times 2068} \times \frac{25.94}{175}$$

$$- 0.00507 \times \frac{400}{0.7 \times 0.95 \times 2068} = 0.000211$$

$$A_f = \rho_f bd = 36.87\ \text{mm}^2/\text{m}\ \ \text{since}\ \varepsilon_c'$$

$$= 0.00182\ \text{which is different from}\ 0.002\quad (\text{close to}\ 37.63\ \text{mm}^2/\text{m})$$

$$\#\ \text{of NSM tapes per meter} = \frac{36.87}{32} = 1.152$$

$$\text{spacing} = \frac{1000}{1.152} = 867.9\ \text{mm}\ \ \text{Use one NSM tape @ 850 mm.}$$

Negative moment section at exterior support:

No. 10 bars @ 250 mm

$$\text{No. of bars/m} = \frac{1000}{250} = 4$$

Area of reinforcement/m = 100 mm^2 × 4 = 400 mm^2/m

$$\rho_s = \frac{A_{s-}}{bd} = \frac{400}{1000 \times 175} = 0.00229$$

$$\omega = \rho_s \frac{f_y}{f_c'} = 0.00229 \times \frac{400}{25} = 0.03657$$

$$R = \omega - 0.59\omega^2 = 0.03578 = \frac{M_n}{f_c' bd^2}$$

$$M_n = 0.03578 \times 25 \times 1000 \times 175^2 = 27{,}394{,}793.5\ \text{N-mm/m} = 27.39\ \text{kN-m/m}$$

$$\bar{M}_u = \frac{\bar{w}_u L_n^2}{24} = \frac{17.884 \times 5.5^2}{24} = 22.54\ \text{kN-m/m}$$

$$\phi M_n = 0.9 \times 27.39 = 24.65\ \text{kN-m/m} > \bar{M}_u = 22.54\ \text{kN-m/m}$$

Do not need strengthening at this section.

5.4.4 Cover Delamination

This is one of the undesirable flexural failure modes of FRP-strengthened beams that takes place due to the shear and normal stress concentrations at the FRP curtailment. This failure mode typically competes with the ductile crushing and debonding failure modes that take place in moderately reinforced or moderately strengthened beams. This is a controlling failure mode that can be delayed by using end-anchorage U-wraps. Many models have been proposed in the literature to accurately predict this failure mode (Roberts 1989; Ziraba et al. 1994; Quantrill, Hollaway, and Thorne 1996; Täljsten 1997; Malek, Saadatmanesh, and Ehsani 1998; El-Mihilmy and Tedesco 2001). However, these models yield inconsistent results, typically due to the empirical nature of these models and due to the fracture nature of the failure mode, which is difficult to capture by simplified stress analysis. It is important to note that ACI 440.2R-08 does not adopt a separate model that designs against this failure mode, but rather uses the same debonding strain as a limit (Figure 5.19).

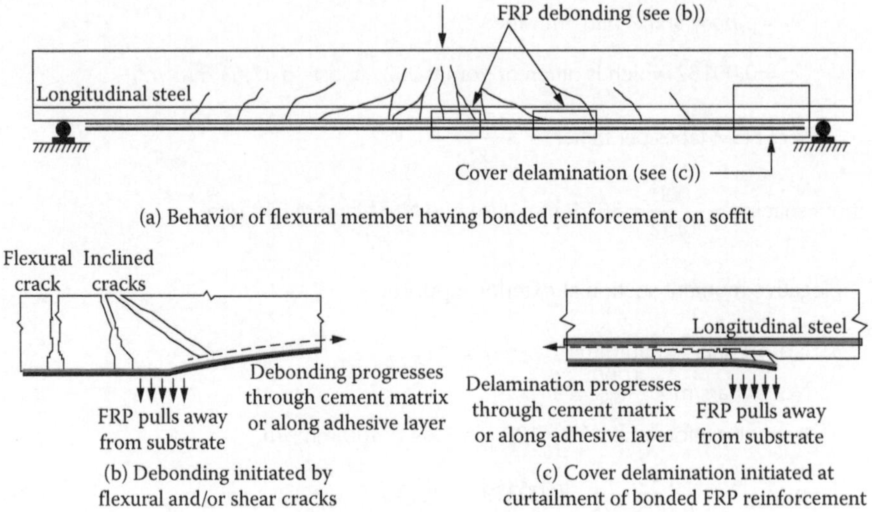

FIGURE 5.19 Cover delamination and debonding failure modes of FRP-strengthened beams. (Courtesy of ACI 440.2R-08.)

Accordingly, the simplified model implemented by FIB Task Group 9.3 FRP (2001) will be studied here.

FIB 2001 presents the ultimate tensile strain approach in the FRP as a means to guarantee FRP anchorage to concrete substrate prior to the cover delamination failure mode. This model was first developed by Holzenkämpfer (1994) and was modified by Neubauer and Rostasy (1997). It is based on bilinear bond-shear-stress–bond-slip law with a linear ascending part and linear descending part, as seen in Figure 5.20.

In this model, the maximum FRP force that can be anchored, $N_{fa,\max}$, and the maximum anchorage length, $l_{b,\max}$, are defined as follows:

$$N_{fa,\max} = \alpha C_1 k_c k_b b \sqrt{E_f t_f f_{ctm}} \quad (5.91)$$

$$l_{b,\max} = \sqrt{\frac{E_f t_f}{C_2 f_{ctm}}} \quad (5.92)$$

where

- α is a reduction factor = 0.9 to account for the effect of inclined cracks on bond strength (Neubauer and Rostasy 1997); $\alpha = 1$ in slabs and beams with sufficient internal and external shear reinforcements;
- k_c is a factor accounting for the state of compaction of concrete, where $k_c = 1.0$ in general for FRP bonded to well-compacted surfaces and $k_c = 0.67$ if FRP is bonded to surface not in contact with the formwork during casting;

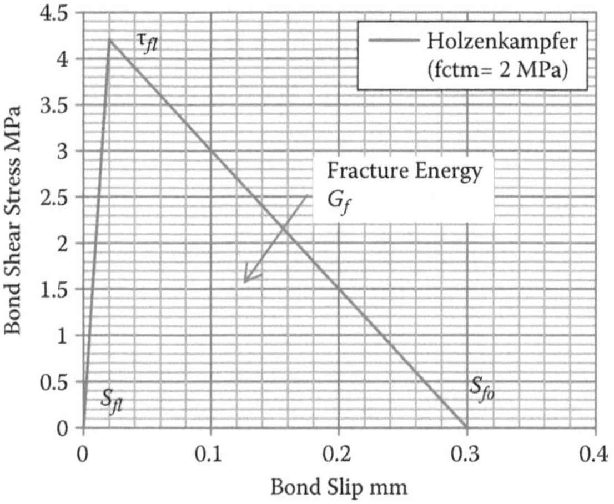

FIGURE 5.20 Holzenkampfer model for bond shear stress-slip of externally bonded FRP to concrete.

k_b is a geometry factor:

$$k_b = 1.06\sqrt{\frac{2-\dfrac{b_f}{b}}{1+\dfrac{b_f}{400}}} \geq 1.0 \quad \text{with} \quad \frac{b_f}{b} \geq \frac{1}{3} \tag{5.93}$$

C_1 and C_2 are coefficients needing to be calibrated from test results:

$$\left.\begin{array}{l} C_1 = 0.64 \\ C_2 = 2 \end{array}\right\} \text{for CFRP strips (plates) and}$$

b, b_f, and t_f are in mm, and E_f and f_{ctm} are in MPa.

For $l_b < l_{b,\max}$, the ultimate force to be anchored is

$$N_{fa} = \left(N_{fa,\max}\right)\frac{l_b}{l_{b,\max}}\left(2-\frac{l_b}{l_{b,\max}}\right) \tag{5.94}$$

According to Equation (4.25) of FIB Task Group 9.3,

$$f_{ctm} \approx \frac{M_{cr}}{bh^2/6}$$

$$\begin{aligned} \text{Thus } f_{ctm} \approx f_r &= 0.62\sqrt{f_c'} \text{ in MPa} \\ &= 7.5\sqrt{f_c'} \text{ in psi} \end{aligned} \tag{5.95}$$

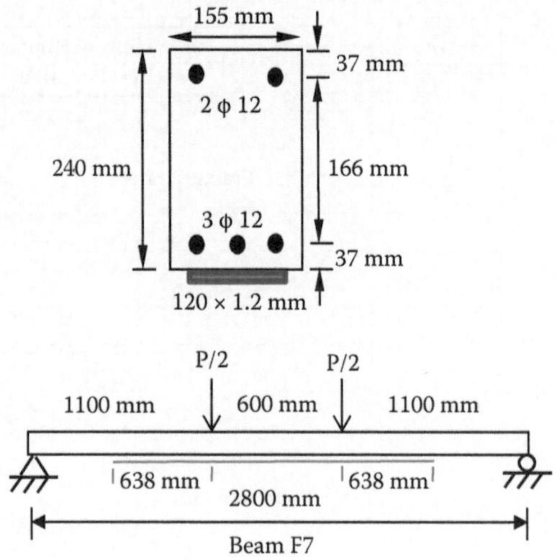

FIGURE 5.21 Example 5.13 showing the beam section and profile.

Example 5.13: Analysis

Fanning and Kelly (2001) tested a series of beams that failed by cover delamination (F5-F10). Beam F7 (Figure 5.21) had the following parameters:

$$f_{fu}^{*} = 2400 \text{ MPa}$$

$$E_f = 155 \text{ GPa}$$

$$A_f = 120 \text{ mm x } 1.2 \text{ mm}$$

$$f_y = 532 \text{ MPa}$$

$$E_s = 204 \text{ GPa}$$

$$E_c = 39.2 \text{ GPa (measured)}$$

$$f_c' = 80 \text{ MPa}$$

$$\frac{L_{af}}{L_a} = \frac{638}{1100} = 0.58$$

Ignore ε_{bi}

Solution:

$$f_{ctm} = f_r = 0.62\sqrt{80} = 5.55 \text{ MPa}$$

$$f_{ct} = 5.0 \text{ MPa (measured)}$$

$$\Rightarrow f_r = \frac{7.5}{6.7} f_{ct} = 5.6 \text{ MPa} \quad \text{O.K.}$$

$$l_{b,\max} = \sqrt{\frac{E_f t_f}{C_2 f_{ctm}}} = \sqrt{\frac{155000 \times 1.2}{2 \times 5.55}} = 129.45 \text{ mm}$$

$$l_b = 638 \text{ mm} > l_{b,\max}$$

$$N_{fa,\max} = \alpha C_1 k_c k_b b \sqrt{E_f t_f f_{ctm}}$$

$$k_b = 1.06\sqrt{\frac{2 - \frac{120}{155}}{1 + \frac{120}{400}}} = 1.03$$

$$N_{fa,\max} = 0.9 \times 0.64 \times 1.0 \times 1.03 \times 155\sqrt{155000 \times 1.2 \times 5.55} = 93431.73 \text{ N} = 93.43 \text{ kN}$$

Calculate the strain at cover delamination:

$$\varepsilon_{fcd} = \frac{N_{fa,\max}}{A_f E_f} = \frac{93.432}{120 \times 1.2 \times 155} = 0.00419$$

$$\varepsilon_{fd} = 0.41\sqrt{\frac{80}{1 \times 155{,}000 \times 1.2}} = 0.0085 >> \varepsilon_{fcd} \text{ Unconservative}$$

$$\alpha f_c' bc + A_s' f_s' = A_s f_y + A_f f_f$$

$$\alpha = \frac{\varepsilon_{cf}}{\varepsilon_c'} - \frac{\varepsilon_{cf}^2}{3\varepsilon_c'^2} = \frac{\bar{\varepsilon}_{fcd} \times c}{(d_f - c) \times \varepsilon_c'} - \frac{\bar{\varepsilon}_{fcd}^2 \times c^2}{3(d_f - c)^2 \varepsilon_c'^2}$$

$$\varepsilon_c' = 1.71\frac{f_c'}{E_c} = 1.71\frac{f_c'}{4700\sqrt{f_c'}} = 3.64 \times 10^{-4}\sqrt{f_c'} = 0.00325 > 0.003 \text{ Crushing strain}$$

Even though the strain $\varepsilon_c' > 0.003$, still use it, since f_c' is very high (80 MPa). Substituting ε_c' and ε_{fcd} into the α equation,

$$\alpha = 1.289\frac{c}{240 - c} - \frac{1.662}{3}\frac{c^2}{(240 - c)^2}$$

$$\left[1.289\frac{c}{240 - c} - \frac{1.662}{3}\frac{c^2}{(240 - c)^2}\right] \times 80 \times 155c + 226 \times 204{,}000 \times \frac{0.00419}{(240 - c)}(c - 37)$$

$$= 339 \times 532 + 93.432 \times 10^3 = 273{,}780$$

Multiply the equation above by $(240 - c)^2$

$$15{,}983.6c^2(240 - c) - 6{,}869.6c^3 + 193{,}175.76 \times (240 - c) \times (c - 37)$$
$$- 273{,}780 \times (240 - c)^2 = 0$$

$$3{,}836{,}064c^2 - 15{,}983.6c^3 - 6{,}869.6c^3 - 193{,}175.76c^2 + 53{,}509{,}685.52c$$
$$- 1{,}715{,}400{,}749 - 1.5769728 \times 10^{10} + 131{,}414{,}400c - 273{,}780c^2 = 0$$

$$-22{,}853.2c^3 + 3{,}369{,}108.24c^2 + 184{,}924{,}085.5c - 1.748512875 \times 10^{10} = 0$$

$$c^3 - 147.42c^2 - 8{,}091.82c + 765{,}106.36 = 0$$

$$c = 57.67 \text{ mm}, \quad \varepsilon_{fd} = 0.41\sqrt{\frac{80}{1 \times 155{,}000 \times 1.2}} = 0.0085 > \varepsilon_{fcd}$$

$$\varepsilon_{cf} = \frac{0.00419}{240 - 57.67} \times 57.67 = 0.00133 < 0.003 \quad \text{No crushing}$$

$$\varepsilon_s = \frac{0.00419}{240 - 57.67}(203 - 57.67) = 0.00334 < 0.005 \text{ Steel in the transition zone}$$

$$\varepsilon_y = \frac{f_y}{E_s} = \frac{532}{204{,}000} = 0.00261 < \varepsilon_s \text{ Tension steel has yielded}$$

$$\bar{\phi} = 0.65 + 0.25\frac{0.00334 - 0.00261}{0.005 - 0.00261} = 0.726$$

$$\varepsilon_s' = \frac{0.00419}{240 - 57.67}(57.67 - 37) = 0.000475 < \varepsilon_y \text{ Compression steel has } \textit{not} \text{ yielded}$$

$$\bar{M}_u = \bar{\phi} A_s f_y (d - \beta c) + \bar{\phi} \Psi_f A_f f_{fe} (d_f - \beta c) + \bar{\phi} A_s' f_s' (\beta c - d')$$

$$\beta = \frac{\frac{1}{3} - \frac{\varepsilon_{cf}}{12\varepsilon_c'}}{1 - \frac{\varepsilon_{cf}}{3\varepsilon_c'}} = \frac{\frac{1}{3} - \frac{0.00133}{12 \times 0.00325}}{1 - \frac{0.00133}{3 \times 0.00325}} = 0.346$$

$$\bar{M}_u = 0.726 \times 339 \times 532 \times (203 - 0.346 \times 57.67) + 0.726 \times 0.85 \times 120 \times 1.2$$
$$\times 155{,}000 \times 0.00419 \times (240 - 0.346 \times 57.67) + 0.726 \times 226 \times 204{,}000$$
$$\times 0.000475 \times (0.346 \times 57.67 - 37)$$
$$= 36{,}394{,}940.42 \text{ N-mm} = 36.4 \text{ kN-m}$$

$$P_{\text{failure\_exp}} = 97.5 \text{ kN} = P_n$$

$$\bar{M}_n = \frac{P_n}{2} \times 1.1 = 53.625 \text{ kN-m}$$

$$\bar{\phi}_{\text{implied}} = \frac{\bar{M}_u}{\bar{M}_n} = 0.679 < 0.726 \text{ Slightly conservative}$$

On the other hand, Beam F10 is identical to Beam F7 except for having a shorter FRP plate, ($L_{af} = l_b = 550$ mm < 638 mm for Beam F7). In this case,

$$l_b = 550 \text{ mm} > l_{b,max} = 129.45 \text{ mm}$$

Accordingly, the rest of the calculations are the same as that of Beam B7 (i.e., $\bar{M}_u = 36.4$ kN-m). But,

$$P_{\text{failure\_exp}} = 82 \text{ kN} = P_n$$

$$\bar{M}_n = \frac{P_n}{2} \times 1.1 = 45.1 \text{ kN-m}$$

$$\bar{\phi}_{\text{implied}} = \frac{\bar{M}_u}{\bar{M}_n} = 0.807 > 0.726 \text{ N.G. Slightly nonconservative}$$

Therefore, the threshold value of the $l_{b,max}$ is too short (nonconservative), and the FIB model in this case does not accurately account for the variation in the FRP plate length within the shear span. Nevertheless, the plate length in the shear span of Beam B10 is very short, $\frac{L_{af}}{L_a} = 0.5$, which is not typically used in practice.

Example 5.14: Design

Arduini, Tommaso, and Nanni (1997) tested a series of beams, one of which (A4) failed by cover delamination when strengthened with three parallel CFRP plates (one layer), as seen in Figure 5.22. The beam had the following parameters:

$\overline{P}_n = 110$ kN $f_y = 540$ MPa

$P_n = 70$ kN $E_s = 200$ GPa

$f'_c = 33$ MPa $E_f = 167$ GPa

$E_c = 25$ GPa $t_f = 1.3$ mm

$f^*_{fu} = 2906$ MPa

Solution:

$$\frac{L_{af}}{L_a} = \frac{550}{700} = 0.786$$

$$\overline{M}_n = \frac{\overline{P}_n}{2} \times 0.7 \text{ m} = 38.5 \text{ kN-m}$$

$$f_{ctm} = f_r = 0.62\sqrt{33} = 3.56 \text{ MPa}$$

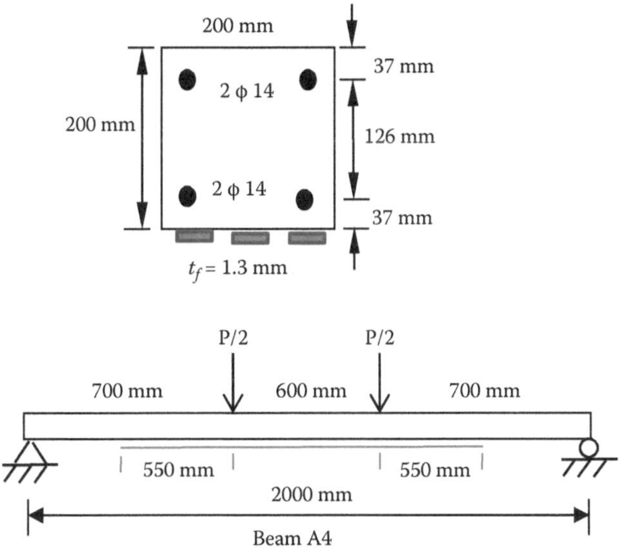

FIGURE 5.22 Example 5.14 showing the beam section and profile.

$$l_{b,\max} = \sqrt{\frac{E_f t_f}{C_2 f_{ctm}}} = \sqrt{\frac{167{,}000 \times 1.3}{2 \times 3.56}} = 174.62 \text{ mm}$$

$$l_b = 550 \; mm \; > l_{b,\max}$$

Since b_f is unknown a priori, assume $k_b = 1.0$ in Equation (5.91).

$$N_{fa,\max} = \alpha C_1 k_c k_b b \sqrt{E_f t_f f_{ctm}}$$

$$N_{fa,\max} = 0.9 \times 0.64 \times 1.0 \times 1.0 \times 200\sqrt{167{,}000 \times 1.3 \times 3.56} = 101{,}276.2 \text{ N} = 101.28 \text{ kN}$$

$$\varepsilon_{fcd} = \frac{N_{fa,\max}}{A_f E_f} = \frac{101.28}{150 \times 1.3 \times 167} = 0.00311 \rightarrow \text{by estimating } A_f$$

This is an obvious disadvantage of this model in solving design problems.

$$\varepsilon_{fd} = 0.41\sqrt{\frac{33}{1 \times 167{,}000 \times 1.3}} = 0.00505$$

$$0.9 C_E \varepsilon_{fu} = 0.9 \times 0.95 \times \frac{2{,}906}{167{,}000} = 0.0149$$

ε_{fcd} controls

Unstrengthened beam capacity:

$$0.85 f_c' b \beta_1 c + A_s' f_s' = A_s f_y$$

$$0.85 \times 33 \times 200 \times 0.826c + 2 \times \frac{\pi}{4} \times 14^2 \times \frac{600}{c}(c - 37) = 2 \times \frac{\pi}{4} \times 14^2 \times 540$$

$$4{,}633.86c^2 + 184{,}725.65c - 6{,}834{,}848.98 - 166{,}253.1c = 0$$

$$c^2 + 3.986c - 1474.98 = 0$$

$$c = \frac{-3.986 + \sqrt{3.986^2 + 4 \times 1474.98}}{2} = 36.46 \text{ mm}$$

$c \approx d' \Rightarrow$ Ignore the effect of compression steel.

$$M_n = A_s f_y \left(d - \frac{\beta_1 c}{2} \right) = 307.9 \times 540 \times \left(163 - \frac{0.826 \times 36.46}{2} \right)$$

$$= 24{,}597{,}727.9 \text{ N-mm} = 24.6 \text{ kN-m}$$

$$M_{n,\exp} = \frac{70}{2} \times 0.7 = 24.5 \text{ kN-m} \quad \text{O.K.}$$

$$\frac{\bar{M}_n}{M_n} = \frac{38.5}{24.6} = 1.565$$

$$\frac{\bar{M}_n}{M_n} = 0.7815 \frac{\rho_f f_f}{\rho_s f_y} \frac{d_f}{d} + 1 = 1.565$$

$$\lambda = \frac{\rho_f f_{fe}}{\rho_s f_y} \frac{d_f}{d} \Rightarrow \rho_f = \frac{0.565 \times 0.009445 \times 540 \times 163}{0.7815 \times 167{,}000 \times 0.00311 \times 200} = 0.00579$$

where

$$\rho_s = \frac{A_s}{bd} = \frac{307.9}{200 \times 163} = 0.009445$$

$$A_f = \rho_f \times bd = 188.75 \text{ mm}^2$$

$$b_f = \frac{A_f}{t_f} = 145.2 \text{ mm} \approx 150 \text{ mm} \quad \text{Actual}$$

Use 3 plates of 50 × 1.3 mm at the soffit of the beam, as done experimentally. Confirm that the failure mode is cover delamination and not concrete crushing.

$$\rho_f = \alpha \frac{f'_c}{f_{fe}} \frac{c}{d} - \rho_s \frac{f_y}{f_{fe}} \quad \text{Ignoring compression steel}$$

$$\alpha = \frac{\varepsilon_{fcd} \times c}{(d_f - c) \times \varepsilon'_c} - \frac{\varepsilon_{fcd}^2 \times c^2}{3(d_f - c)^2 \varepsilon_c'^2}$$

$$0.00579 = \left(\frac{0.00311c}{(200 - c) \times 0.00209} - \frac{0.00311^2 c^2}{3(200 - c)^2 \times 0.00209^2} \right) \times \frac{33}{519.4} \times \frac{c}{163}$$

$$- 0.009445 \times \frac{540}{519.4}$$

where

$$\varepsilon'_c = 1.71 \frac{f'_c}{4{,}700\sqrt{f'_c}} = 0.00209$$

$$f_{fe} = 167{,}000 \times 0.00311 = 519.4 \text{ MPa}$$

$$0.0156 = \left[1.488 \times \frac{c}{200 - c} - 0.738 \times \frac{c^2}{(200 - c)^2} \right] \times 0.0003898c$$

Multiply the previous equation by $(200 - c)^2$:

$$0.0156(40000 - 400c + c^2) = 0.00058c^2(200 - c) - 0.000288c^3$$

$$0.000868c^3 - 0.1004c^2 - 6.24c + 624 = 0$$

$$c^3 - 115.67c^2 - 7188.94c + 718894 = 0$$

$$c = 69.07 \text{ mm}$$

$$\varepsilon_{cf} = \frac{0.00311 \times 69.07}{(200 - 69.07)} = 0.00164 < 0.003 \; O.K.$$

$$\varepsilon_s = \frac{0.00311 \times (163 - c)}{(200 - c)} = 0.00223 < \frac{540}{200000} = 0.0027 \quad \text{No yielding of tension steel}$$

$$f_s = 200000 \times 0.00223 = 446.23 \text{ MPa}$$

$$\varepsilon_s' = \frac{0.00311 \times (c - 37)}{(200 - c)} = 0.00076 < \frac{540}{200000} = 0.0027 \quad \text{No yielding of top steel}$$

$$f_s' = 200000 \times 0.00076 = 152 \text{ MPa}$$

$$\beta = \frac{\frac{1}{3} - \frac{0.00164}{12 \times 0.00209}}{1 - \frac{0.00164}{3 \times 0.00209}} = 0.363$$

$$\frac{\bar{M}_u}{\bar{\phi}} = \frac{\bar{M}_u}{0.65} = \bar{M}_n = 38.5 \times 10^6 N - mm = A_s f_s (d - \beta c) + \Psi_f A_f f_{fe} (d_f - \beta c) + A_s' f_s' (\beta c - d')$$

$$38.5 \times 10^6 = 307.9 \times 446.23 \times (163 - 0.363 \times 69.07) + 0.85 \times A_f \times 519.4$$
$$\times (200 - 0.363 \times 69.07) + 307.9 \times 152 \times (0.363 \times 69.07 - 37)$$

$$A_f = \frac{20107767.52}{77228.78} = 260.4 \text{ mm}^2 >> 195 \text{ mm}^2 \quad \text{(actual)}$$

Iterate one more time:

$$\alpha f_c' bc + A_s' f_s' = A_s f_s + A_f f_{fcd}$$

$$\alpha = \frac{\varepsilon_{fcd} \times c}{(d_f - c) \times \varepsilon_c'} - \frac{\varepsilon_{fcd}^2 \times c^2}{3(d_f - c)^2 \varepsilon_c'^2}$$

$$\left(1.488 \times \frac{c}{200 - c} - 0.738 \times \frac{c^2}{(200 - c)^2}\right) \times 33 \times 200c + 307.9 \times 622 \times \frac{(c - 37)}{(200 - c)}$$
$$= 307.9 \times 622 \times \frac{(163 - c)}{(200 - c)} + 260.4 \times 519.4$$

$c = 65.06$ mm

$\varepsilon_{cf} = 0.0015 < 0.003$ O.K.

$\varepsilon_s = 0.00226 < 0.0027$ No yielding of tension steel

$$f_s = 200000 \times 0.00226 = 451.5 \text{ MPa}$$

$$\varepsilon_s' = \frac{0.00311 \times (c - 37)}{(200 - c)} = 0.000647 < 0.0027 \quad \text{No yielding of top steel}$$

$$f_s' = 200{,}000 \times 0.000647 = 129.34 \text{ MPa}$$

$$\beta = \frac{\frac{1}{3} - \frac{0.0015}{12 \times 0.00209}}{1 - \frac{0.0015}{3 \times 0.00209}} = 0.36$$

$$\frac{\bar{M}_u}{\bar{\phi}} = \frac{\bar{M}_u}{0.65} = \bar{M}_n = 38.5 \times 10^6 = A_s f_s (d - \beta c) + \Psi_f A_f f_{fe} (d_f - \beta c) + A_s' f_s' (\beta c - d')$$

$$38.5 \times 10^6 = 307.9 \times 451.5 \times (163 - 0.36 \times 65.06) + 0.85 \times A_f \times 519.4$$
$$\times (200 - 0.36 \times 65.06) + 307.9 \times 129.34 \times (0.36 \times 65.06 - 37)$$

$$A_f = \frac{19{,}636{,}993.8}{77{,}957.6} = 251.9 \text{ mm}^2 \approx 260.4 \text{mm}^2 >> 195 \text{ mm}^2 \quad \text{(actual)}$$

No need to iterate.

$$b_f = \frac{251.9}{1.3} = 193.8 \text{ mm} \approx 200 \text{mm}$$

Use four plates of 50 × 1.3 mm covering the entire soffit of beam.
It can be concluded that the FIB model is very conservative in this case. It is also worth mentioning that the experimental results presented by Arduini, Tommaso, and Nanni (1997) indicate that Beam A4 does not undergo tensile steel yielding, which is in agreement with the design calculations shown here.

5.4.5 FRP Debonding

This is the fifth flexural failure mode in beams strengthened with FRP. The debonding is initiated at one of the flexural or shear cracks along the span, as seen in Figure 5.19b. Accordingly, it is referred to as intermediate induced cracking. It is a dominant failure mode in moderately reinforced, moderately strengthened beams with FRP sheets or plates extending close to the support competing with the ductile crushing failure. This failure mode may be avoided in two ways: (a) by keeping the maximum FRP strain below the strain of FRP debonding as specified by ACI 440.2R-08, (b) by anchoring the beam's flexural FRP by transverse U-wraps designed according to the adapted shear-friction model of ACI 318-11 (Rasheed, Larson, and Peterman 2006; Rasheed et al. 2010, 2011). A different model limits the interface shear stress after cracking to a limiting value (Rasheed, Larson, and Nayyeri Amiri 2013). In this section, the first approach of limiting the FRP maximum strain is explored in comparison with existing experimental results. This limiting strain is defined by ACI 440.2R-08 to be

$$\varepsilon_{fd} = 0.083\sqrt{\frac{f_c'}{nE_f t_f}} \quad \text{in U.S. units}$$
$$= 0.41\sqrt{\frac{f_c'}{nE_f t_f}} \quad \text{in SI units} \qquad (5.96)$$

The limiting FRP ratio between concrete crushing and FRP debonding may be written as

$$a_b^{\max} = \beta_1 \frac{\varepsilon_{cu} d_f}{\varepsilon_{cu} + \bar{\varepsilon}_{fd}} \qquad (5.97)$$

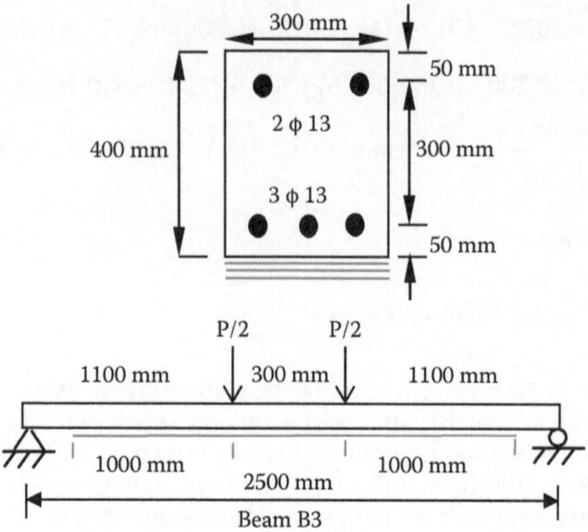

FIGURE 5.23 Example 5.15 showing the beam section and profile tested by Arduini, Tommaso, and Nanni (1997).

$$\rho_f^{max} = 0.85 \frac{f_c'}{f_{fd}} \frac{a_b^{max}}{d} - \rho_s \frac{f_y}{f_{fd}} \tag{5.98}$$

$$f_{fd} = E_f \varepsilon_{fd} \tag{5.99}$$

In design problems, the statistically correlated linear equation may be used to determine a quick initial value for ρ_f by substituting f_{fd} into Equation (5.86) to replace f_f. Furthermore, a single iteration may be needed to adjust ρ_f to its final value by invoking the force and moment equilibrium:

Example 5.15: Analysis

Arduini, Tommaso, and Nanni (1997) tested a series of beams strengthened with CFRP sheets, one of which failed by sheet debonding (Beam B3). The beam (Figure 5.23) had the following parameters:

$A_s = 398$ mm^2 $\quad E_f = 400$ GPa

$A_s' = 265$ mm^2 $\quad f_{fu}^* = 3000$ MPa

$f_y = 340$ MPa $\quad b_f = 300$ mm

$f_c' = 30$ MPa $\quad t_f = 0.17$ mm

No. of layers $(n) = 3$

$M_{n,exp} = 51.66$ kN-m $\quad (P_n = 93.93$ kN$)$

$\bar{M}_{n,exp} = 126.69$ kN-m $\quad (\bar{P}_n = 230.35$ kN$)$

Solution:

$\frac{L_{af}}{L_a} = \frac{1000}{1100} = 0.91 > 0.9$ Plate debonding is more likely to occur than cover delamination (Rasheed, Larson, and Nayyeri Amiri 2013).

$$\varepsilon_{fd} = 0.41\sqrt{\frac{30}{3 \times 400{,}000 \times 0.17}} = 0.00497 < 0.9C_E\varepsilon_{fu} = 0.00641$$

Determine the failure mode (whether ductile crushing or debonding):

$$\beta_1 = 0.85$$

$$a_b^{max} = \beta_1 \frac{\varepsilon_{cu} \times d_f}{\varepsilon_{cu} + \varepsilon_{fd}} = 0.85 \times \frac{0.003 \times 400}{0.003 + 0.00497} = 127.95 \text{ mm}$$

assuming $\varepsilon_{bi} = 0$, since it is a research specimen (strengthening happened under self-weight).

$$\rho_f^{max} = 0.85 \frac{f_c'}{f_{fd}} \frac{a_b^{max}}{d} - \rho_s \frac{f_y}{f_{fd}}$$

$$f_{fd} = 400{,}000 \times 0.00497 = 1{,}988 \text{ MPa}$$

$$\rho_f^{max} = 0.85 \times \frac{30}{1988} \times \frac{127.95}{350} - \frac{398}{300 \times 350} \times \frac{340}{1988} = 0.00404$$

$$\rho_f = \frac{3 \times 300 \times 0.17}{300 \times 350} = 0.001457 < \rho_f^{max} \text{ debonding controls}$$

$$\alpha f_c' bc + A_s' f_s' = A_s f_y + A_f f_{fd}$$

$$\alpha = \frac{\varepsilon_{cf}}{\varepsilon_c'} - \frac{1}{3}\left(\frac{\varepsilon_{cf}}{\varepsilon_c'}\right)^2 = 2.50\frac{c}{400 - c} - \frac{1}{3}\left(2.50\frac{c}{400 - c}\right)^2$$

$$\text{where } \varepsilon_{cf} = \frac{\varepsilon_{fd} c}{d_f - c} = 0.00497\frac{c}{400 - c}$$

$$\varepsilon_c' = 1.71\frac{f_c'}{4700\sqrt{f_c'}} = 3.64 \times 10^{-4}\sqrt{f_c'} = 0.00199$$

$$\varepsilon_s' = \frac{\varepsilon_{fd}(c - d')}{d_f - c} = 0.00497\frac{c - 50}{400 - c}$$

Substituting these expressions into the force equilibrium equation,

$$\left[2.50 \times \frac{c}{400 - c} - 2.1 \times \frac{c^2}{(400 - c)^2}\right] \times 30 \times 300c + 265 \times 200000 \times 0.00497\frac{c - 50}{400 - c}$$

$$= 398 \times 340 + 153 \times 1988$$

Multiply this equation by $(400 - c)^2$,

$$22{,}590c^2(400-c)-18{,}900c^3+263{,}410(400-c)(c-50)=439{,}484(400-c)^2$$

$$9{,}036{,}000c^2-22{,}590c^3-18{,}900c^3+118{,}534{,}500c-263{,}410c^2-5{,}268{,}200{,}000$$

$$-7.031744\times10^{10}+351{,}587{,}200c-439{,}484c^2=0$$

$$-41{,}490c^3+8{,}333{,}106c^2+470{,}121{,}700c-7.558564\times10^{10}=0$$

$$c=85.912\text{ mm}$$

$$\varepsilon_{cf}=0.00497\times\frac{85.912}{400-85.912}=0.00136<0.003\quad\text{O.K.}$$

$$\varepsilon_s=0.00497\times\frac{350-85.912}{400-85.912}=0.00418<0.005\text{ Tension steel is in the transition zone.}$$

$$\bar{\phi}=0.65+0.25\frac{0.00418-0.0017}{0.005-0.0017}=0.838$$

$$\text{where}\quad\varepsilon_y=\frac{f_y}{E_s}=\frac{340}{200{,}000}=0.0017$$

$$\varepsilon_s'=0.00497\times\frac{85.912-50}{400-85.912}=0.00057<\varepsilon_y\quad\text{No yielding of compression steel}$$

$$f_s'=E_s\varepsilon_s'=200{,}000\times0.00057=114\text{ MPa}$$

$$\beta=\frac{\frac{1}{3}-\frac{0.00136}{12\times0.00199}}{1-\frac{0.00136}{3\times0.00199}}=0.358$$

$$\bar{M}_u=\bar{\phi}A_sf_y(d-\beta c)+\bar{\phi}\Psi_fA_ff_{fd}(d_f-\beta c)+\bar{\phi}A_s'f_s'(\beta c-d')$$

$$\bar{M}_u=0.838\times398\times340(350-0.358\times85.912)+0.838\times0.85\times153\times1988$$

$$\times(400-0.358\times85.912)+0.838\times265\times114\times(0.358\times85.912-50)$$

$$=115{,}713{,}284.7\text{ N-mm}=115.713\text{ kN-m}$$

$$\bar{M}_{n,\exp}=126.69\text{ kN-m}$$

$$\bar{\phi}_{\text{implied}}=\frac{\bar{M}_u}{\bar{M}_{n,\exp}}=0.913>0.838\quad\text{Slightly unconservative, should be}<0.838.$$

Using the statistically correlated Equation (5.86),

$$\lambda=\frac{\rho_ff_{fd}}{\rho_sf_y}\frac{d_f}{d}=\frac{0.001457\times1988}{0.00379\times340}\times\frac{400}{350}=2.569$$

$$\frac{\bar{M}_n}{M_n}=0.7815\times2.569+1=3\Rightarrow\bar{M}_n=3\times48.365=145.47>\bar{M}_{n,\exp}=126.69$$

Where M_n = 48.365 kN-m from Example 5.11, the prediction of equation (5.96) is slightly unconservative.

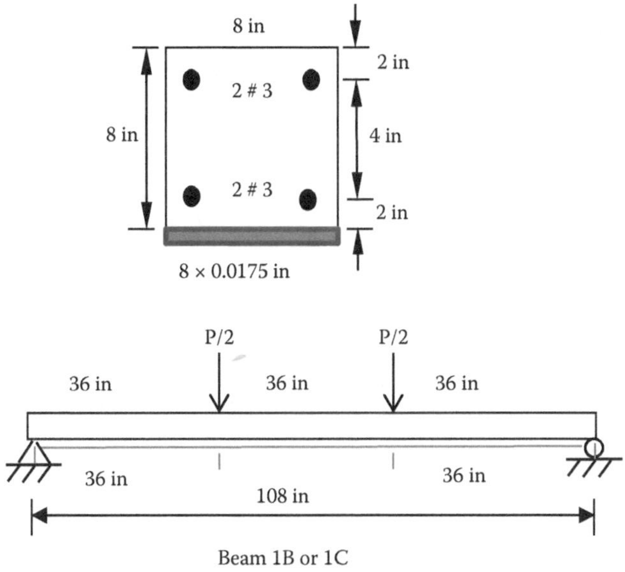

FIGURE 5.24 Example 5.16 showing the beam section and profile.

Example 5.16: Design

Ross et al. (1999) tested a series of beams by fixing the area of FRP and changing the area of steel. Beams 1B and 1C in group 1 failed by FRP debonding and are selected in this example (Figure 5.24).

$$f_c' = 7950 \text{ psi}$$

$$f_y = 60 \text{ ksi}$$

$$A_f = 8 \text{ in.} \times 0.0175 \text{ in.} \quad \text{(actual)}$$

$$L_f = 108 \text{ in.}$$

$$f_{fu}^* = 320 \text{ ksi}$$

$$E_f = 20000 \text{ ksi}$$

$$A_s = 0.22 \text{ in.}^2$$

$$E_s = 29{,}000 \text{ ksi}$$

$$\left.\begin{array}{l}\bar{P}_n(1B) = 80.1 \text{ kN} = 18 \text{ kips}\\ \bar{P}_n(1C) = 71.2 \text{ kN} = 16 \text{ kips}\end{array}\right\} \text{use the more conservative ultimate load}$$

$$\bar{M}_n = \frac{16}{2} \times 3 \text{ ft} = 24 \text{ k-ft}$$

$$\varepsilon_{bi} = 0 \text{ (research specimen)}$$

Solution:

Assume ductile crushing failure mode.

$$Q_1 = \rho_s \frac{f_y}{f_c'} \frac{d}{d_f} = \frac{0.22}{8 \times 6} \times \frac{60}{7.95} \times \frac{6}{8} = 0.0259$$

$$Q_2 = \frac{\bar{M}_u}{\phi f_c' b d_f^2} + Q_1\left(\Psi_f - \frac{d}{d_f}\right) = \frac{24 \times 12}{7.95 \times 8 \times 8^2} + 0.0259 \times \left(0.85 - \frac{6}{8}\right) = 0.0733$$

$$\left(\frac{a}{d_f}\right)^2 - (2 - 0.208 \times 0.0259) \times \left(\frac{a}{d_f}\right) + 2.77 \times 0.0733 = 0$$

$$\frac{a}{d_f} = \frac{1.995 - \sqrt{1.995^2 - 11.08 \times 0.0733}}{2} = 0.1078 \Rightarrow a = 0.862\ in. \Rightarrow \beta_1$$
$$= 0.6525 \Rightarrow c = 1.321 \text{ in.}$$

$$\varepsilon_{fe} = \frac{0.003}{c}(8 - c) - \varepsilon_{bi} = \frac{0.003}{1.321}(8 - 1.321) - 0 = 0.01516$$

$$\varepsilon_{fd} = 0.083\sqrt{\frac{f_c'}{nE_f t_f}} = 0.083\sqrt{\frac{7{,}950}{1 \times 20{,}000 \times 0.0175}} = 0.0125 < 0.9\varepsilon_{fu}$$

$$0.9\varepsilon_{fu} = 0.9 \times C_E \times \frac{f_{fu}}{E_f} = 0.01368 \quad \text{debonding controls}$$

Using the statistically correlated linear equation for initial estimate of ρ_f:
Unstrengthened beam capacity:

$$0.85 f_c' b \beta_1 c + A_s' f_s' = A_s f_y$$

$$0.85 \times 7.95 \times 8 \times 0.6525c + 0.22 \times \frac{87}{c}(c - 2) = 0.22 \times 60$$

$$35.274c^2 + 19.14c - 38.28 = 13.2c$$

$$35.274c^2 + 5.94c - 38.28 = 0$$

$$c = \frac{-5.94 + \sqrt{5.94^2 + 4 \times 35.274 \times 38.28}}{2 \times 35.274} = 0.961 \text{ in.}$$

$$\varepsilon_s' = \frac{0.003}{0.961}(0.961 - 2) = -0.00324 \text{ (tension)} > \varepsilon_y$$

$$f_s' = f_y$$

$$0.85 f_c' b \beta_1 c = A_s' f_y + A_s f_y$$

$$c = \frac{0.44 \times 60}{0.85 \times 7.95 \times 8 \times 0.6525} = 0.748 \text{ in.}$$

$$M_n = A_s f_y\left(d - \frac{\beta_1 c}{2}\right) - A_s' fy\left(\frac{\beta_1 c}{2} - d'\right)$$

$$M_n = 0.22\times 60\left(6-\frac{0.6525\times 0.748}{2}\right)+0.22\times 60\times\left(2-\frac{0.6525\times 0.748}{2}\right) = 99.157\text{ k-in.}$$

$$M_n = 8.263\text{ k-ft}$$

$$M_{n,\exp} = \frac{P_n}{2}\times 3\text{ ft} = \frac{6}{2}\times 3 = 9\text{ k-ft}$$

Close values, use the M_n this time:

$$\frac{\bar{M}_{n,\exp}}{M_n} = \frac{24}{8.263} = 2.9$$

$$\frac{\bar{M}_{n,\exp}}{M_n} = 0.7815\lambda + 1 = 2.9$$

$$\lambda = 2.431$$

$$\rho_f = \frac{\lambda\rho_s f_y d}{f_{fd} d_f} = \frac{2.431\times 0.00458\times 60\times 6}{20000\times 0.0125\times 8} = 0.00201$$

$$\rho_{f,\text{actual}} = \frac{8\times 0.0175}{8\times 6} = 0.00292 \quad \text{unconservative}$$

Perform another iteration:

$$\rho_f = \alpha\frac{f'_c}{f_{fd}}\frac{c}{d} - \rho_s\frac{f_y}{f_{fd}}$$

$$\alpha = \frac{\varepsilon_{fd}\times c}{(8-c)\times\varepsilon'_c} - \frac{\varepsilon_{fd}^2\times c^2}{3(8-c)^2\times\varepsilon_c'^2} = 4.682\frac{c}{8-c} - 7.306\frac{c^2}{(8-c)^2}$$

$$\varepsilon'_c = 1.71\frac{f'_c}{57{,}000\sqrt{f'_c}} = 3\times 10^{-5}\sqrt{f'_c} = 0.00267$$

$$0.00201 = \left(4.682\times\frac{c}{(8-c)} - 7.306\times\frac{c^2}{(8-c)^2}\right)\frac{7.95}{250}\times\frac{c}{6} - 0.00458\times\frac{60}{250}$$

$$0.00311\times(8-c)^2 = 0.02481c^2\times(8-c) - 0.03872c^3$$

$$0.06353c^3 - 0.1954c^2 - 0.04976c + 0.199 = 0$$

$$c = 1.069\text{ in.}$$

$$\varepsilon_{cf} = \frac{0.0125\times 1.069}{(8-1.069)} = 0.001928$$

$$\beta = \frac{\frac{1}{3} - \frac{0.001928}{12\times 0.00267}}{1 - \frac{0.001928}{3\times 0.00267}} = 0.36$$

$$\varepsilon_s = \frac{0.0125(6-c)}{(8-c)} = 0.00889 > 0.005 \Rightarrow \bar{\phi} = 0.9$$

$$\varepsilon'_s = \frac{0.0125}{(8-c)}(c-2) = -0.00168$$

Compression steel is in tension (not yielded),

$$f_s' = 29{,}000 \times 0.00168 = 48.72 \text{ ksi}$$

$$\frac{\bar{M}_u}{\phi} = \bar{M}_n = 24 \times 12 = A_s f_y (d - \beta c) + \Psi_f A_f f_{fd} (d_f - \beta c) + A_s' f_s' (d' - \beta c)$$

$$288 = 0.22 \times 60 \times (6 - 0.36 \times 1.069) + 0.85 A_f \times 250 \times (8 - 0.36 \times 1.069)$$
$$+ 0.22 \times 48.72 \times (2 - 0.36 \times 1.069)$$

$$A_f = \frac{196.57}{0.85 \times 250 \times (8 - 0.36 \times 1.069)} = 0.121 \text{ in.}^2 < 0.14 \text{ in.}^2 \quad \text{Unconservative}$$

This is the second example in which the FRP debonding model shows unconservative results.

Chapter Problems

Problem 5.1

A library building has a simply supported rectangular concrete beam reinforced with four No. 6 bars in tension, two No. 3 bars in compression and No. 3 stirrups at 6 in. (152 mm) on center in shear. The details of the beam are shown in Figure 5.P.1. As part of the library upgrade, the beam is subjected to a 40% increase in live load, as shown in Figure 5.P.2. Determine the CFRP area that needs to be bonded in flexure. Use VWRAP CFRP sheets with a modulus of 33,000 ksi (227,527 MPa) and a tensile strength of 550 ksi (3,792 MPa) based on the net fiber area that has a net fiber thickness of 0.0065 in. (0.165 mm).

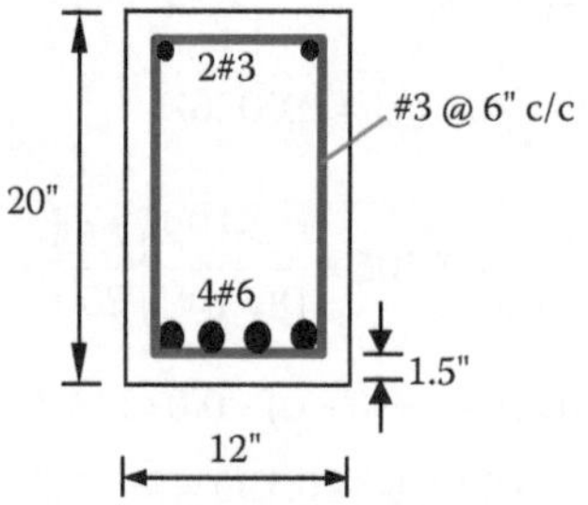

FIGURE 5.P.1

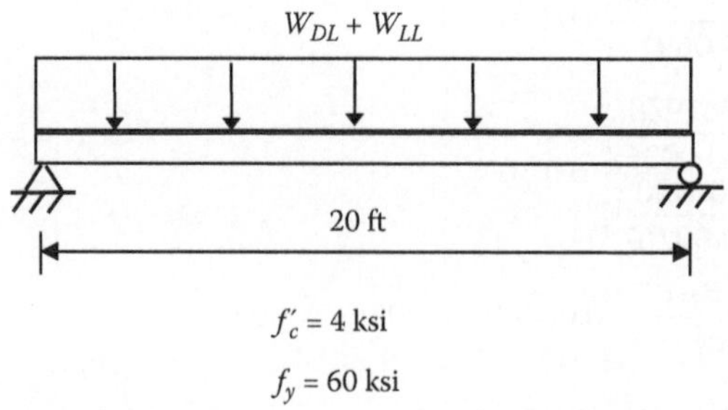

FIGURE 5.P.2

Span	20 ft	6.1 m
B	12 in.	305 mm
H	20 in.	508 mm
f'_c	4 ksi	27.6 MPa
f_y	60 ksi	414 MPa
Top bars	#3	$\phi = 9.5$ mm
Main bars	#6	$\phi = 19$ mm
W_{DL}	0.9 k/ft	13.14 kN/m
W_{LL}	0.9 k/ft	13.14 kN/m

Problem 5.2

Repeat Example 5.1 to design the required FRP by taking $d_f = h + \frac{t}{2}$, $t_f = 0.04$ in. with no rounding off or approximating numbers in the equations. Also, check if the strengthening limits due to loss of composite action are satisfied.

Problem 5.3

For the beam section given in Figure 5.P.3, design for the area of FRP required to increase the section capacity by 4% *only*.

$$f'_c = 4\ ksi, \quad f_y = 60\ ksi, \quad \varepsilon_{bi} = 0.0008$$

$$d_t = d = 14 - 1 - \frac{3}{8} - \frac{1}{2} \times \frac{7}{8} = 12.19"$$

$$t_f = 0.04", \quad E_f = 5260 \text{ ksi}, \quad f_{fu} = 90 \text{ ksi}$$

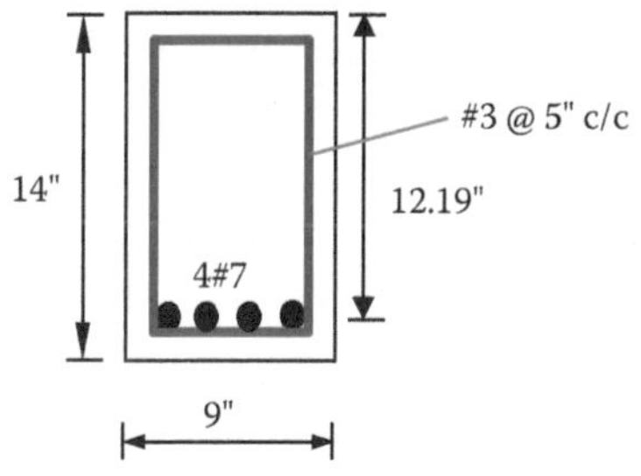

FIGURE 5.P.3

Problem 5.4

Derive the coefficients of the approximate Equation (5.77) A_3, B_3, and D_3 for a doubly reinforced rectangular section in case of yielding and no yielding of compression steel for the failure mode of FRP rupture.

Problem 5.5

Decker (2007) designed, built, strengthened, and tested rectangular reinforced concrete beams having the following properties:

b	6 in. (152.4 mm)
h	12 in. (304.8 mm)
C_t	1 in. (25.4 mm)
C_b	1 in. (25.4 mm)
h_s	10 in. (254 mm)
b_s	4 in. (101.6 mm)

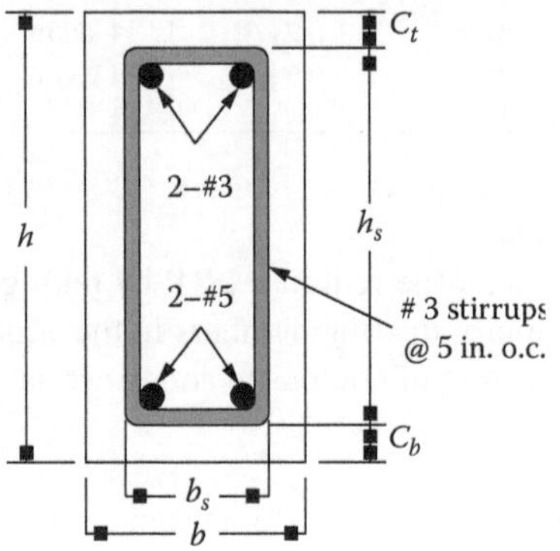

FIGURE 5.P.4

See Figure 5.P.4.

$f_c' = 5.2$ ksi (35.9 MPa), $f_y = 69.64$ ksi (480.2 MPa), $\varepsilon_{bi} = 0$

$t_f = 0.0065"$ (0.165 mm), $E_f = 33{,}000$ ksi (227.5 GPa), $f_{fu} = 550$ ksi (3792 MPa)

Decker used VWRAP C100 CFRP for strengthening. Use the same material to strengthen the beam to a 20% increase in the unstrengthened beam moment capacity. Assume no need to strengthen the beam in shear.

Problem 5.6

Repeat Problem 5.5 with a 40% strengthening ratio increase over the bare beam. Designate the failure mode involved. Assume no need to strengthen the beam in shear.

Problem 5.7

Repeat Problem 5.5 with a 60% strengthening ratio increase over the bare beam. Designate the failure mode involved. Assume fire not to be a hazard in this case. Also, assume no need to strengthen the beam in shear.

Problem 5.8

Decker (2007) designed, built, strengthened, and tested reinforced concrete T-beams having the following properties:

b_f	16 in. (406.4 mm)
h_f	4 in. (101.6 mm)
b_w	6 in. (152.4 mm)
h_w	8 in. (203.2 mm)

C_t	1 in. (25.4 mm)
C_b	1 in. (25.4 mm)
h_s	10 in. (254 mm)
b'	13 in. (330.2 mm)
b_s	4 in. (101.6 mm)

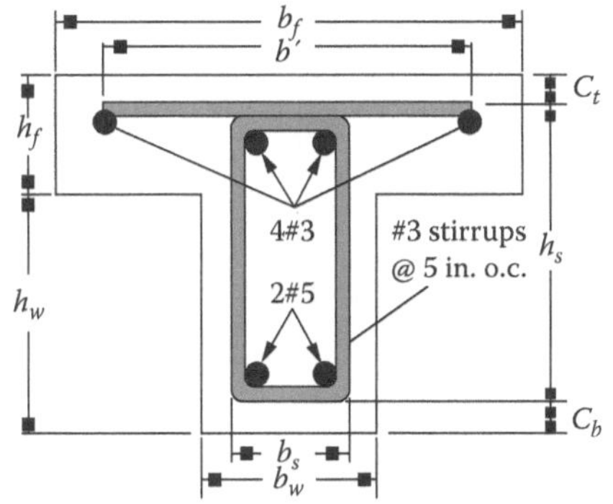

FIGURE 5.P.5

See Figure 5.P.5.

$f_c' = 4.9$ ksi (33.5 MPa), $f_y = 69.64$ ksi (480.2 MPa), $\varepsilon_{bi} = 0$

$t_f = 0.0065"$ (0.165 mm), $E_f = 33{,}000$ ksi (227.5 GPa), $f_{fu} = 550$ ksi (3792 MPa)

Decker used VWRAP C100 CFRP for strengthening. Use the same material to strengthen the beam to a 30% increase in the unstrengthened beam moment capacity. Designate the failure mode involved. Assume no need to strengthen the beam in shear.

Problem 5.9

Repeat Problem 5.8 with a 60% strengthening ratio increase over the bare beam. Designate the failure mode involved. Assume fire not to be a hazard in this case. Also, assume no need to strengthen the beam in shear.

Problem 5.10

Design the slab given in Figure 5.P.6 based on ACI 318-11 code, and then strengthen it based on ACI 440.2R-08, assuming no change in dead load. The slab is a two-span, continuous solid, one-way slab with the following parameters:

$w_{LL} = 4.8$ kN/m^2
$w_{LLupgrade} = 7.0$ kN/m^2
$f_y = 414$ MPa
$f_c' = 25$ MPa
Slab thickness = 180 mm

Use NSM tape from Example 5.12 for negative moment region and CFRP sheets from Example 5.15 for positive moment region.

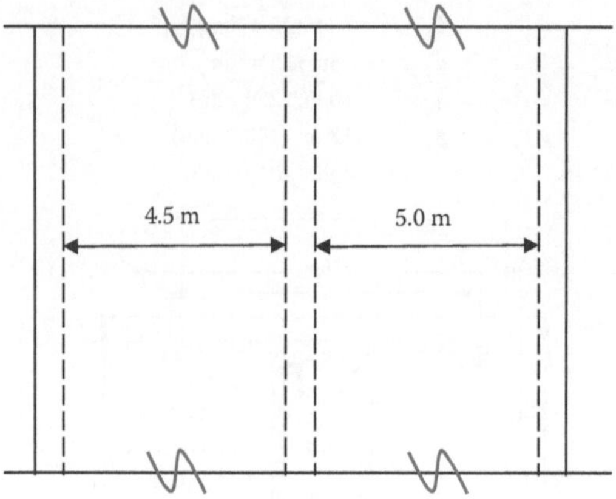

FIGURE 5.P.6

Problem 5.11

Al-Tamimi et al. (2011) have designed, built, strengthened, and tested the beam shown in Figure 5.P.7 (B85P). It has a CFRP plate extending 1,521 mm along the span of the beam covering 85% of the shear span. ReCon Plate CFL Grade CF210, 100-mm wide by 1.4-mm thick, was used to strengthen the beam (f^*_{fu} = 2500 MPa, E_f = 215 GPa). The average measured compressive strength of the concrete was 54 MPa. The modulus of elasticity and yield strength for the tensile steel reinforcement were 202 GPa and 611 MPa, respectively. Determine the capacity of the beam using the FIB model for cover delamination. Compare the ultimate load predicted to the experimental ultimate load at cover delamination of 60.7 kN and conclude.

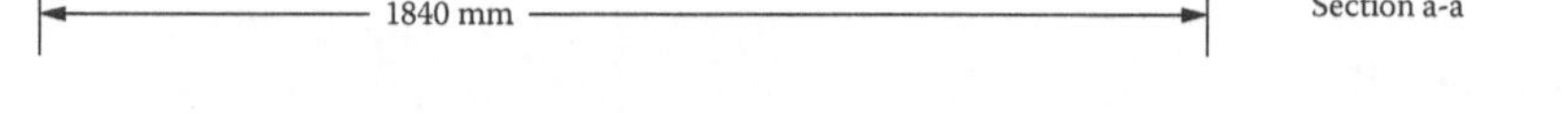

(a) Longitudinal and cross-section reinforcement details of the beam

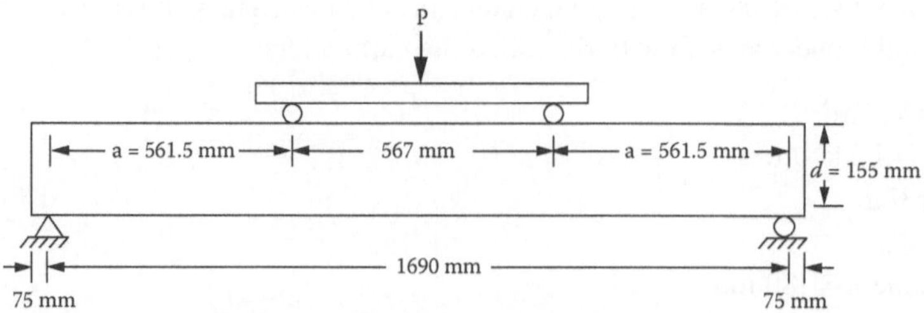

(b) Details of Tested Beams

FIGURE 5.P.7

Problem 5.12

Al-Tamimi et al. (2011) have designed, built, strengthened, and tested the beam shown in Figure 5.P.8 (B70PW). It has a CFRP plate extending 1352 mm along the span of the beam covering 70% of the shear span. ReCon Plate CFL Grade CF210, 100-mm wide by 1.4-mm thick, was used to strengthen the beam (f^*_{fu} = 2500 MPa, E_f = 215 GPa). The average measured compressive strength of the concrete was 54 MPa. The modulus of elasticity and yield strength for the tensile steel reinforcement were 202 GPa and 611 MPa, respectively. Determine the capacity of the beam using the ACI440.2R-08 model for plate debonding. Compare the ultimate load predicted to the experimental ultimate load at plate debonding of 67.7 kN and conclude.

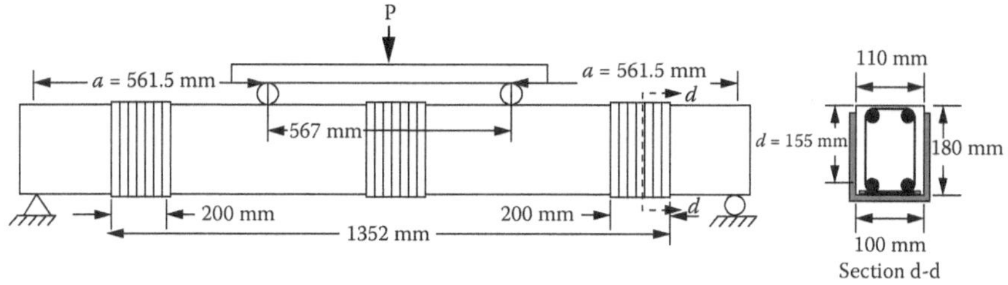

FIGURE 5.P.8

Problem 5.13

Wuertz (2013) designed, built, strengthened, and tested rectangular reinforced concrete beams having the section geometry shown in Figure 5.P.9:

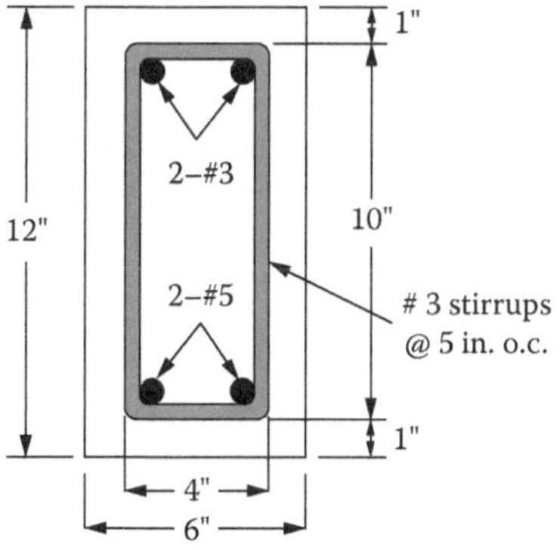

FIGURE 5.P.9

$f'_c = 9.4$ ksi (64.9 MPa), $f_y = 71$ ksi (490 MPa), $\varepsilon_{bi} = 0$

$t_f = 0.05"$ (1.27 mm), $E_f = 3030$ ksi (20.9 GPa), $f_{fu} = 66.72$ ksi (460.3 MPa)

Wuertz used V-WRAP EG50 GFRP for strengthening in addition to two NSM steel bars. The two NSM bars were installed in grooves in the concrete cover. The GFRP was applied as one layer only and was wrapped 5" (127 mm) up the sides from the soffit. Use the same strengthening design to determine the strengthened beam moment capacity if the GFRP was continuously anchored against cover delamination and sheet debonding. Assume no need to strengthen the beam in shear.

Problem 5.14

Spadea, Bencardino, and Swamy (1998) designed, built, strengthened, and tested a rectangular reinforced concrete beam (A3.3) having the section geometry and beam profile shown in Figures 5.P.10 and 5.P.11. Knowing that beam A3.3 had distributed steel U-wrap anchors along its span, it is expected that the two modes of cover delamination and plate debonding were avoided. Considering the rest of the possible failure modes, design the FRP plate width (b_f) so that it will provide an ultimate load of 98.3 kN. Compare the answer to the 80-mm width of the actual plate.

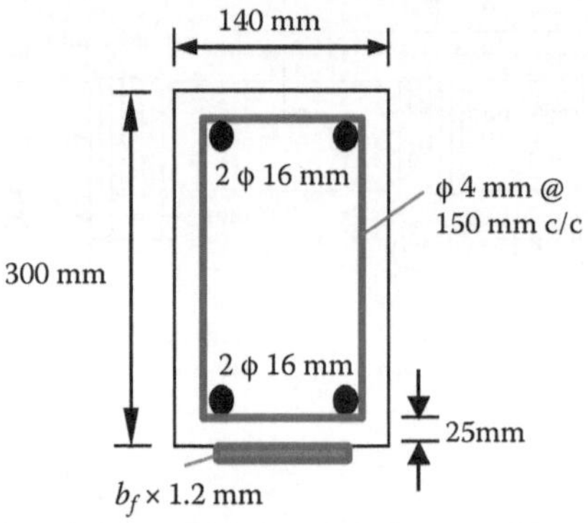

FIGURE 5.P.10

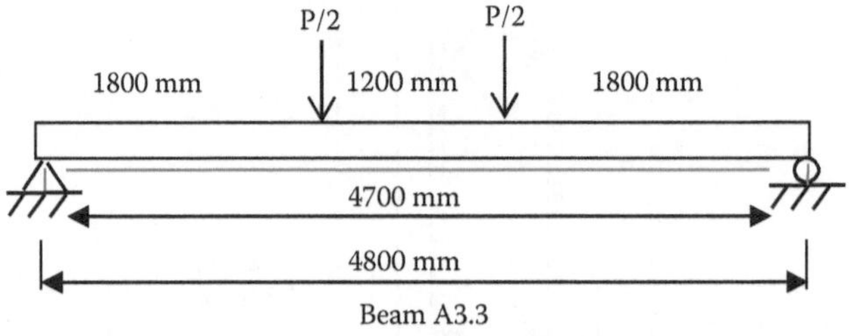

FIGURE 5.P.11

$f_c' = 30.5$ MPa, $f_y = 435$ MPa, $\varepsilon_{bi} = 0$

$t_f = 1.2$ mm, $E_f = 152$ GPa, $f_{fu} = 2300$ MPa

REFERENCES

ACI 318-05. 2005. Building code requirements for structural concrete and commentary. ACI Committee 318, Farmington Hills, MI.

ACI 318-11. 2011. Building code requirements for structural concrete and commentary. ACI Committee 318, Farmington Hills, MI.

ACI 440.2R-08. 2008. Guide for the design and construction of externally bonded FRP systems for strengthening concrete structures. ACI Committee 440, Farmington Hills, MI.

Alkhrdaji, T., A. Nanni, and R. Mayo. 2000. Upgrading Missouri transportation infrastructure: Solid RC decks strengthened with FRP systems. *Transportation Research Record* 1740:157–69.

Al-Tamimi, A. K., R. Hawileh, J. Abdalla, and H. A. Rasheed. 2011. Effects of ratio of CFRP plate length to shear span and end anchorage on flexural behavior of SCC R/C beams. *Journal of Composites for Construction* 15 (6): 908–19.

Arduini, M., A. Tommaso, and A. Nanni. 1997. Brittle failure in FRP plate and sheet bonded beams. *ACI Structural Journal* 94 (4): 363–70.

Chaallal, O., M.-J. Nollet, and D. Perraton. 1998. Strengthening of reinforced concrete beams with externally bonded fiber-reinforced plastic plates: Design guidelines for shear and flexure. *Canadian Journal of Civil Engineering* 25:692–704.

Decker, B. R. 2007. A method of strengthening monitored deficient bridges. M.S. thesis, Kansas State Univ., Manhattan, KS.

El-Mihilmy, M., and J. Tedesco. 2001. Prediction of anchorage failure for reinforced concrete beams strengthened with fiber-reinforced polymer plates. *ACI Structural Journal* 98 (3): 301–14.

Fanning, P., and O. Kelly. 2001. Ultimate response of RC beams strengthened with CFRP plates. *Journal of Composites for Construction* 5 (2): 122–27.

FIB Task Group 9.3 FRP. 2001. Externally bonded FRP reinforcement for RC structures. Technical report committee 9.3, Federation Internationale Du Beton.

GangaRao, H. V. S., and P. V. Vijay. 1998. Bending behavior of concrete beams wrapped with carbon fabric. *ASCE Journal of Structural Engineering* 124 (1): 3–10.

Holzenkämpfer, O. 1994. Ingenieurmodelle des Verbundes geklebter Bewehrung für Betonbauteile. PhD diss.

Malek, A., H. Saadatmanesh, and M. Ehsani. 1998. Prediction of failure load of R/C beams strengthened with FRP plate due to stress concentration at the plate end. *ACI Structural Journal* 95 (2): 142–52.

Meier U. 1987. Repair of bridges with high performance composite materials. *Material und Technik* 15 (4): 125–28.

Neubauer, U., and F. S. Rostasy. 1997. Design aspects of concrete structures strengthened with externally bonded CFRP-plates. In *Proceedings of The Seventh International Conference on Structural Faults and Repair*, ed. M. C. Forde. Vol. 2, Concrete and composites, 109–18. Edinburgh, Scotland.

Okeil, A. M., Y. Bingol, and T. Alkhrdaji. 2007. Analyzing model uncertainties for concrete beams flexurally strengthened with FRP laminates. In *Proceedings of the Transportation Research Board 86th Annual Meeting*, Jan. 21-25, 2007, Washington, DC.

Quantrill, R., L. Hollaway, and A. Thorne. 1996. Predictions of the maximum plate end stresses of FRP strengthened beams: Part II. *Mag. Concrete Res.* 48 (177): 343–51.

Rasheed, H. A., R. R. Harrison, R. J. Peterman, and T. Alkhrdaji. 2010. Ductile strengthening using externally bonded and near surface mounted composite systems. *Composite Structures* 92 (10): 2379–90.

Rasheed, H. A., K. H. Larson, and S. Nayyeri Amiri. 2013. Analytical solution of interface shear stresses in FRP strengthened concrete beams. *ASCE Journal of Engineering Mechanics* 139 (1): 18–28.

Rasheed, H. A., K. H. Larson, and R. J. Peterman. 2006. Analysis and design procedure for FRP-strengthened prestressed concrete T-girders considering strength and fatigue. *Journal of Composites for Construction* 10 (5): 419–32.

Rasheed, H. A., and N. H. Motto. 2010. Strength design equations for FRP strengthened concrete beams. *Australian Journal of Structural Engineering* 11 (2): 103–16.

Rasheed, H. A., M. Nassajy, S. Al-Subaie, S. M. Abrishamchian, and A. Al-Tamimi. 2011. Suppressing delamination failure mode in concrete beams strengthened with short CFRP laminates. *Mechanics of Advanced Materials and Structures* 18 (3): 194–200.

Rasheed, H. A., and S. Pervaiz. 2003. Closed form design equations for flexural strengthening of RC beams. *Composites Part B: Engineering* 34 (6): 539–50.

Roberts, T. 1989. Approximate analysis of shear and normal stress concentrations in the adhesive layer of plated RC beams. *Structural Engineer* 67 (12): 228–33.

Ross, C. A., D. M. Jerome, J. W. Tedesco, and M. L. Hughes. 1999. Strengthening of reinforced concrete beams with externally bonded composite laminates. *ACI Structural Journal* 96 (2): 212–20.

Saqan, E., H. A. Rasheed, and R. A. Hawileh. 2013. An efficient design procedure for flexural strengthening of RC beams based on ACI 440.2R-08. *Composites Part B: Engineering* 49 (June): 71–79.

Spadea, G., B. Bencardino, and R. N. Swamy. 1998. Structural behavior of composite RC beams with externally bonded CFRP. *Journal of Composites for Construction* 2 (3): 132–37.

Täljsten, B. 1997. Strengthening of beams by plate bonding. *J. Mater. Civ. Eng.* 9 (4): 206–12.

Triantafillou, T. C., and N. Plevris. 1992. Strengthening of RC beams with epoxy-bonded fibre-composite materials. *Materials and Structures* 25:201–11.

Wuertz, A. F. 2013. Strengthening rectangular beams with NSM steel bars and externally bonded GFRP. M.S. thesis, Kansas State Univ., Manhattan, KS.

Ziraba, Y., M. Baluch, I. Basunbul, and A. Sharif. 1994. Guidelines toward the design of reinforced concrete beams with external plates. *ACI Structural Journal* 91 (6): 639–46.

6 Shear Strengthening of Concrete Members

6.1 OVERVIEW

Experimental studies have shown that shear strengthening of concrete members is possible to accomplish by applying the fibers transverse to the member axis or perpendicular to the shear cracks (Triantafillou 1998; Triantafillou and Antonopoulos 2000). Increasing the shear strength may cause the flexural failure to dominate the behavior, which generally provides more ductile response than that of shear failure (ACI 440.2R-08 2008).

Chapter 11 of ACI 440.2R-08 provides a model to compute the extra shear strength furnished by FRP to concrete beams and columns. This extra shear strength depends on the concrete strength, the type of FRP wrapping scheme, the geometry of the concrete member, and the amount of steel shear reinforcement provided. For external FRP shear stirrups in the form of discrete strips, the center-to-center spacing between the strips should not exceed $d/4$ plus the width of the strip. In other words, the clear distance between the strips should not exceed $d/4$.

6.2 WRAPPING SCHEMES

For beams and columns of rectangular sections, the FRP wrapping schemes are as illustrated in Figure 6.1.

1. *Complete wrapping*: FRP systems wrapped around all four sides of the section represent the most efficient scheme. This scheme is typically used in columns, since all four sides of the section are accessible (Figure 6.1a).
2. *U-wrapping*: FRP systems wrapped around three sides of the section in beams are used to improve shear strength where it is impractical to completely wrap the section due to the existence of the slab attached to the beam from both sides (interior beams, Figure 6.1b) or from one side (exterior or spandrel beams, Figure 6.1c). This technique is less efficient than the complete wrapping scheme.
3. *Side bonding*: FRP systems bonded along the two opposite sides of the web of the beam are used to improve the shear strength where it is impractical to U-wrap the section due to the existence of section enlargement such as the bulb tee (Figure 6.1d). This technique is the least efficient of the three schemes.

For all three wrapping schemes, it is possible to install a continuous sheet along the span of the member as well as install discrete strips. The first case of fully

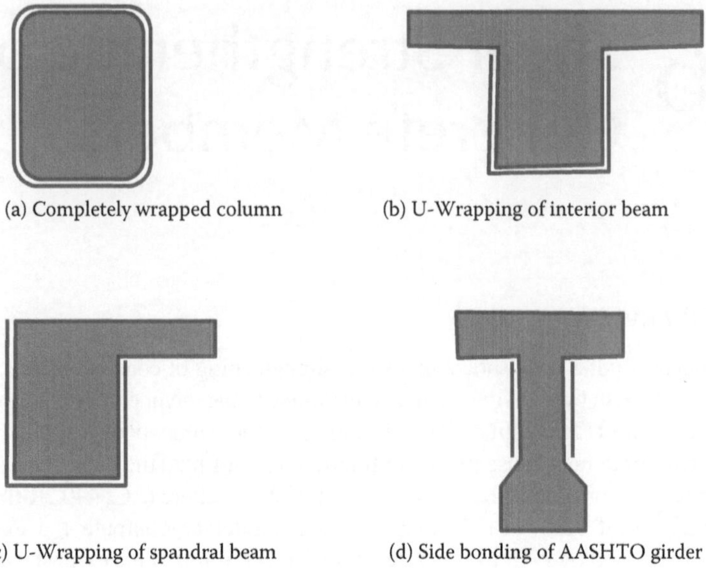

FIGURE 6.1 Wrapping schemes for FRP shear strengthening.

encasing the member with FRP is discouraged due to the potential prevention of migration of moisture as well as the more difficult detection of potential localized delaminations. The second case of installing discrete strips is the more widely used technique, in which the center-to-center distance between strips depends on the amount of FRP needed, which translates into the total number of strip layers and the width of each strip. The wider the strips and the fewer the number of layers used, the more efficient the shear strengthening scheme is, since the interface shear stress transfer to concrete is lower, which makes the failure mode of sheet debonding less likely to take place.

6.3 ULTIMATE AND NOMINAL SHEAR STRENGTH

The design shear strength of the member having FRP shear strengthening should exceed the required shear or the demanded shear value.

$$\bar{\phi}\bar{V}_n \geq \bar{V}_u \tag{6.1}$$

The strength reduction factor $\bar{\phi}$ is as per ACI 318-05 (2005). The load factors used to compute $\bar{V}_u$ are also per ACI 318-05, as specified by ACI 440.2R-08. These factors are the same as those of ACI 318-11 (2011).

The nominal shear strength, including the FRP contribution, may be obtained by adding this FRP contribution to that of concrete and reinforcing steel stirrups as follows:

$$\bar{\phi}\bar{V}_n = \bar{\phi}(V_c + V_s + \Psi_f V_f) \tag{6.2}$$

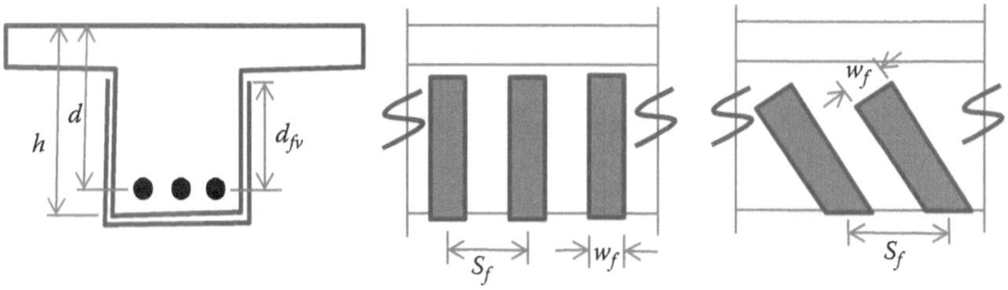

FIGURE 6.2 Variables for FRP shear-strengthening design.

where V_c is computed based on Equations (11.3) to (11.8) of ACI 318-05, V_s is computed based on Section 11.5.7.2 of ACI 318-05 as specified by ACI 440.2R-08. The value of Ψ_f is determined to be 0.85 for U-wraps and side bonding based on a reliability study of existing experimental data. On the other hand, Ψ_f is selected to be 0.95 for fully wrapped section based on the lower bond dependence. V_f is computed based on the following equations:

$$V_f = \frac{A_{fv} f_{fe} (\sin\alpha + \cos\alpha) d_{fv}}{S_f} \tag{6.3}$$

where

$$A_{fv} = 2nt_f w_f \tag{6.4}$$

$$f_{fe} = E_f \varepsilon_{fe} \tag{6.5}$$

where ε_{fe} is the effective tensile strain developed in the FRP shear stirrups at nominal shear strength. This effective tensile strain is controlled by the failure of the strengthened concrete member and by the failure of the FRP stirrups. The variables d_{fv} and S_f are denoted in Figure 6.2.

6.4 DETERMINATION OF ε_{fe}

The value of ε_{fe} is determined as follows:

1. *Fully wrapped section*: The loss of aggregate interlock of the concrete is observed to take place at a fiber strain much lower than the ultimate fiber strain (Priestley, Seible, and Calvi 1996):

$$\varepsilon_{fe} = 0.004 \le 0.75\varepsilon_{fu} \tag{6.6}$$

2. *U-wraps or plies bonded to two sides*: U-wraps (three-sided wraps) and bonded side plies (two-sided strips) have been observed to delaminate from concrete prior to the loss of aggregate interlock. Accordingly, bond

stresses have been analyzed to determine the effective strain level that can be attained (Triantafillou 1998):

$$\varepsilon_{fe} = k_v \varepsilon_{fu} \leq 0.004 \tag{6.7}$$

The bond reduction coefficient (k_v) is a function of the concrete strength, the ply stiffness, and the wrapping scheme. This coefficient may be computed based on the model by Khalifa et al. (1998):

$$k_v = \frac{k_1 k_2 L_e}{468\varepsilon_{fu}} \leq 0.75 \quad \text{in U.S. customary units} \tag{6.8}$$

$$k_v = \frac{k_1 k_2 L_e}{11{,}900\varepsilon_{fu}} \leq 0.75 \quad \text{in SI units} \tag{6.9}$$

where

$$k_1 = \left(\frac{f_c'}{4000}\right)^{2/3} \quad \text{in lb-in. units} \tag{6.10}$$

$$k_1 = \left(\frac{f_c'}{27}\right)^{2/3} \quad \text{in SI units} \tag{6.11}$$

$$k_2 = \begin{cases} \dfrac{d_{fv} - L_e}{d_{fv}} & \text{for U-wraps} \\ \dfrac{d_{fv} - 2L_e}{d_{fv}} & \text{for two bonded sides} \end{cases} \tag{6.12}$$

ACI 440.2R-08 defines L_e as the active bond length, which is the length over which the majority of the bond stress is maintained:

$$L_e = \frac{2500}{\left(nt_f E_f\right)^{0.58}} \quad \text{in lb-in. units} \tag{6.13}$$

$$L_e = \frac{23{,}300}{\left(nt_f E_f\right)^{0.58}} \quad \text{in SI units} \tag{6.14}$$

Even though the k_v factor has been validated for a high-shear, low-moment region, ACI 440.2R-08 reports that it is sufficiently conservative for areas of high flexural and shear stresses as well as for negative-moment regions.

6.5 REINFORCEMENT LIMITS

The shear force provided by steel stirrups and FRP stirrups is limited to the level set by ACI 318-05 for the force in steel only to avoid crushing of the concrete struts.

$$V_s + V_f \le 8\sqrt{f_c'}b_w d \qquad \text{in lb-in. units} \tag{6.15}$$

$$V_s + V_f \le 0.66\sqrt{f_c'}b_w d \qquad \text{in SI units} \tag{6.16}$$

Example 6.1: Design

For the beam in Example 2.3, design the shear strengthening required if $\bar{w}_u = 7$ k/ft (see Figure 6.3).

$f_c' = 4$ ksi $\qquad f_{yt} = 50$ ksi

$t_f = 0.0065"$ $\qquad E_f = 33{,}000$ ksi $\qquad f_{fu}^* = 550$ ksi

Solution:

The existing shear reinforcement per Example 2.3 is:

No. 4 stirrups at 4.5" o.c. for the first 20.8" from the support
No. 4 stirrups at 6" o.c. up to 9.9 ft from the support
No. 4 stirrups at 9" o.c. up to 12 ft from the support

$$\bar{V}_u = \bar{w}_u \frac{L_n}{2} = 7 \times \frac{28}{2} = 98 \text{ k}$$

$$\bar{V}_{ud} = 98 - 7 \times \frac{19.75}{12} = 86.48 \text{ k}$$

$\bar{\phi} V_c = 22.48$ k (from Example 2.3)

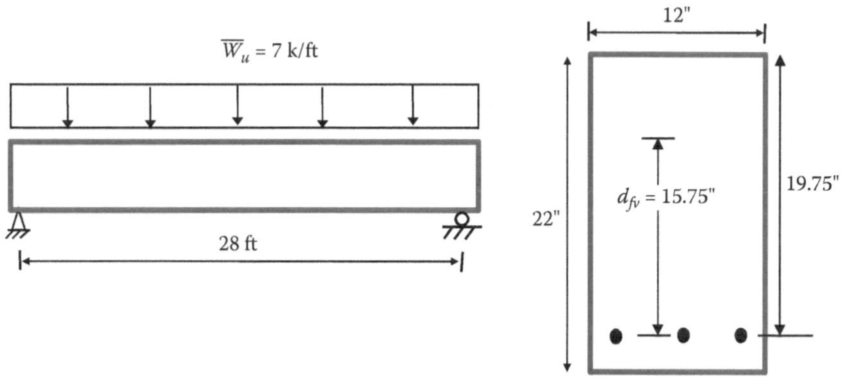

FIGURE 6.3 Beam profile and cross section for Example 6.1; $d_{fv} = 15.75"$

No. 4 stirrups at 4.5" o.c. for a distance of 20.8"

$$\bar{\phi}V_{s1} = \frac{0.75 \times 0.4 \times 50 \times 19.75}{4.5} = 65.83 \text{ k}$$

$\bar{\phi}V_c + \bar{\phi}V_{s1} = 88.31 > \bar{V}_{ud}$ No need for external shear strengthening in this region.

No. 4 stirrups at 6" up to 9.9'

$$\bar{\phi}V_{s2} = \frac{0.75 \times 0.4 \times 50 \times 19.75}{6} = 49.375 \text{ k}$$

$$\bar{\phi}V_c + \bar{\phi}V_{s2} = 71.86 \text{ k}$$

Determine x_{cs2} distance from support to $\left(\bar{V}_u = \bar{\phi}(V_c + V_{s2})\right)$:

$$\bar{\phi}(V_c + V_s) = \bar{V}_u - \bar{w}_u x_{cs2} \Rightarrow x_{cs2} = \frac{98 - 71.86}{7} = 3.73 \text{ ft}$$

$$\bar{V}_{u20.8"} = 98 - 7 \times \frac{20.8"}{12} = 85.87 \text{ k}$$

$$\bar{\phi}\Psi_f V_f = 85.87 - 71.86 = 14.01 \text{ k}$$

$$V_f = \frac{14.01}{0.75 \times 0.85} = 21.98 \text{ k}$$

$$= \frac{A_{fv} f_{fe} d_{fv}}{S_f}$$

$$V_f = \frac{2 \times 0.0065 \times w_f \times E_f \times k_v \varepsilon_{fu} \times 15.75}{\left(w_f + \dfrac{d}{4}\right)} = 21.98 \text{ k}$$

$$\varepsilon_{fu} = C_E \varepsilon_{fu}^* = 0.95 \times \frac{550}{33{,}000} = 0.0158$$

$$k_1 = 1$$

$$L_e = \frac{2{,}500}{\left(1 \times 0.0065 \times 33{,}000 \times 10^3\right)^{0.58}} = 2.022 \text{ in.}$$

$$k_2 = \frac{15.75 - 2.022}{15.75} = 0.872$$

$$k_v = \frac{1 \times 0.872 \times 2.022}{468 \times 0.0158} = 0.238 < 0.75$$

$$21.98\left(w_f + \frac{19.75}{4}\right) = 25.41 w_f$$

$$w_f = \frac{108.53}{3.43} = 31.64 \text{ in.} \qquad \text{Too big}$$

$$21.98(w_f + 2) = 25.41 w_f$$

$$w_f = \frac{43.96}{3.43} = 12.82 \text{ in.} \qquad \text{Use 13" @ 15" c/c (one layer)}$$

Theoretically between 1.73 ft (20.8″) and 3.73 ft
Practically between 1.73 ft and 9.9 ft

$$\bar{\phi}V_{s3} = \frac{0.75 \times 0.4 \times 50 \times 19.75}{9} = 32.92 \text{ k}$$

$$\bar{\phi}(V_c + V_{s3}) = 22.48 + 32.92 = 55.4 \text{ k}$$

Determine x_{cs3} *distance from support to* $\bar{V}_u = \bar{\phi}(V_c + V_{s3})$

$$\bar{\phi}(V_c + V_{s3}) = 55.4 = \bar{V}_u - \bar{w}_u x_{cs3}$$

$$x_{cs3} = \frac{98 - 55.4}{7} = 6.09' \text{ outside } 9.9'\text{-}11.96' \text{ where steel stirrup spacing} = 6''$$

(i.e., no need for FRP between 9.9 ft and 12 ft)

$$\bar{V}_u - \bar{w}_u x_c = 98 - 7 \times 9.9 = 28.7k < \bar{\phi}(V_c + V_{s3}) = 55.4 \text{ k} \qquad \text{O.K.}$$

$$\bar{V}_{u11.96} = 98 - 7 \times 12 = 14 \text{ k}$$

$$x_{c4} = \frac{\bar{V}_u - \bar{\phi}V_c/2}{\bar{w}_u} = \frac{98 - 11.24}{7} = 12.39 \text{ ft}$$

Between 12' and 12.4' we need FRP

Use 1 layer of 13" @ 18" c/c for this distance ($d/4$ clear spacing)

Practically use one layer of 13" @ 18" o.c. between 9.9 ft and 12.9 ft (three U-wraps within 3 ft).
Check the reinforcement limit:

$$V_{s2} = \frac{0.4 \times 50 \times 19.75}{6} = 65.83 \text{ k}$$

$$V_f = \frac{2 \times 1 \times 0.0065 \times 13 \times 33{,}000 \times 0.238 \times 0.0158 \times 15.75}{15} = 22.02 \text{ k}$$

$$V_{s2} + V_f = 87.85\text{k} < 8\sqrt{4{,}000} \times 12 \times 19.75 = 119{,}913.57 \text{ lb} = 119.913 \text{ k} \qquad \text{O.K.}$$

Example 6.2: Design

The column in Example 2.4 is located in a building that underwent a change in its importance category from Regular Building to Essential Facility. The loads were increased as follows:

$M_u = 84.38$ k-ft	$M_u = 84.38$ k-ft
$V_u = 18.75$ k	$V_u = 18.75$ k
$P_u = 187.5$ k	$P_u = 24$ k

Figure 6.4 shows the column section and profile.

$f_c' = 4$ ksi $\qquad f_{yt} = 40$ ksi

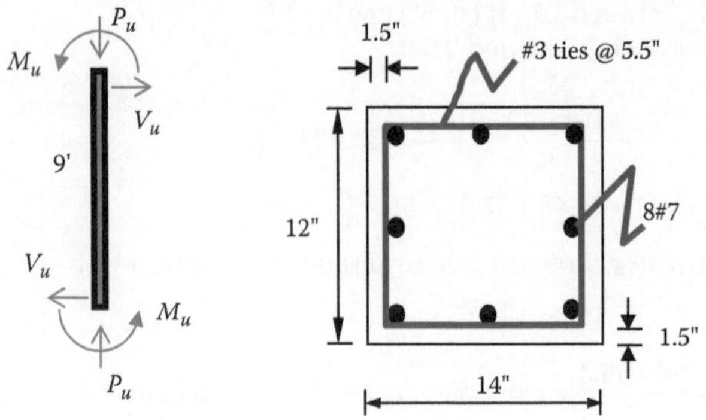

FIGURE 6.4 Example 6.2 showing the column section and profile.

Solution:

Case A: $N_u = P_u = 187.5$ k

$$\phi V_c = 0.75 \times 2 \times \left[1 + \frac{187{,}500}{2{,}000(14 \times 12)}\right]\sqrt{4{,}000}\,(12)(11.69) = 20{,}734.54 \text{ lb} = 20.73 \text{ k}$$

$$\phi V_c > V_u = 18.75 \text{ k}$$

$$\text{But } V_u > \phi\frac{V_c}{2} = 10.37 \text{ k} \quad \text{Minimum reinforcement is provided.}$$

Case B: $N_u = P_u = 24$ k

$$\phi V_c = 0.75 \times 2 \times \left[1 + \frac{24{,}000}{2{,}000(14 \times 12)}\right]\sqrt{4{,}000}\,(12)(11.69) = 14{,}258.7 \text{ lb} = 14.26 \text{ k}$$

$$\phi V_c < 18.75 \text{ k}$$

$$\phi(V_c + V_s) = 14.26 + \frac{0.75 \times 0.22 \times 40 \times 11.69}{5.5} = 28.29 \text{ k} > 18.75 \text{ k}$$

No external shear strengthening is required. This is an atypical situation, since shear demand is not significantly increased in columns. However, an AASHTO (American Association of State Highway and Transportation Officials) extreme load event on bridge piers may demand high-impact lateral force on the column, which translates into external shear strengthening.

Example 6.3: Analysis

Norris, Saadatmanesh, and Ehsani (1997) tested three beams deficient in shear, two of which were externally strengthened in shear using CFRP. The control beam was designated by C48, and the strengthened beams were designated by IE and IIE. The strengthened beams had two layers of CFRP U-wrap transverse fabric

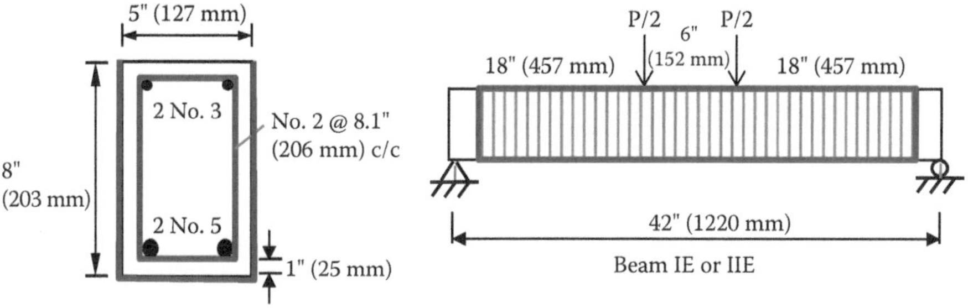

FIGURE 6.5 Example 6.3 showing the beam section and profile.

covering the entire sides of the beams. The geometric and material parameters of the beams are given in Figure 6.5.

Beam IE: $f_{fu} = 56.5$ ksi (389.8 MPa), $E_f = 4943$ ksi (34.1 GPa), $t_f = 0.043$ in. (1.09 mm)
Beam IIE: $f_{fu} = 57.3$ ksi (395.3 MPa), $E_f = 4841$ ksi (33.4 GPa), $t_f = 0.043$ in. (1.09 mm)
$f'_c = 5.3$ ksi (36.6 MPa), $f_y = 61$ ksi (421 MPa)

Solution:

The unstrengthened beams have their steel stirrups provided not in accordance with ACI 318-11 (spacing = 8.1 in. > $d/2$).

$$d = 8 - 1 - \frac{1}{4} - \frac{1}{2} \times \frac{5}{8} = 6.44 \text{ in.} = d_{fv}$$

$$V_c = 2\sqrt{f'_c}b_w d = 2\sqrt{5300} \times 5 \times 6.44 = 4688.4 \text{ lb} = 4.69 \text{ k}$$

No. 2 stirrups at 8.1" o.c.

$$V_s = \frac{0.1 \times 61 \times 6.44}{8.1} = 4.85 \text{ k}$$

$$V_n = V_c + V_s = 9.54\,k < V_{\exp} = 11.8 \text{ k}$$

$$\Psi_f V_f = 0.85 \times \frac{A_{fv} f_{fe} d_{fv}}{S_f}$$

$$\varepsilon_{fu} = C_E \varepsilon^*_{fu} = 0.95 \times \frac{56.5}{4943} = 0.011$$

$$k_1 = \left(\frac{5300}{4000}\right)^{2/3} = 1.21$$

$$L_e = \frac{2500}{\left(1 \times 0.043 \times 4943 \times 10^3\right)^{0.58}} = 2.032 \text{ in.}$$

$$k_2 = \frac{6.44 - 2.032}{6.44} = 0.684$$

$$k_v = \frac{1.21 \times 0.684 \times 2.032}{468 \times 0.011} = 0.327 < 0.75$$

$$\Psi_f V_f = 0.85 \times \frac{2 \times 0.043 \times 12 \times 4943 \times 0.327 \times 0.011 \times 6.44}{12} = 8.37 \text{ k}$$

$$\bar{V}_n = V_c + V_s + \psi_f V_f = 9.54\ k + 8.37\ k = 17.91\ k > \bar{V}_{exp} = 15.28 \text{ k} \qquad \text{N.G.}$$

The ACI 440 model overestimates the shear capacity with FRP in this case. One main reason could be the size effect. The model of Khalifa et al. (1998) was calibrated against typical beam cross sections, while this beam is on the smaller side.

Example 6.4: Analysis

Al-Sulaimani et al. (1994) tested 16 reinforced concrete beams deficient in shear and strengthened in flexure and/or shear using GFRP plates. The control beam was designated as CO, and one of the strengthened beams was designated as SO. The strengthened beam had side-bonded transverse strips that were 20-mm wide at 50-mm on center covering the entire sides of the shear spans. The geometric and material parameters of the strengthened beam are given in Figure 6.6.

$f'c = 37.7$ MPa, $fy = 450$ MPa
GFRP plate: $f_{fu} = 200$ MPa, $E_f = 15.65$ GPa, $\varepsilon_{fu} = 0.01278$, $t_f = 3$ mm

Solution:

The unstrengthened beams have their steel stirrups provided not in accordance with ACI 318-11 (spacing = 200 mm > $d/2$).

$$d = 76 + 37 = 113 \text{ mm} = d_{fv}$$

$$V_c = \frac{\sqrt{f_c'}}{6} b_w d = \frac{\sqrt{37.7}}{6} \times 150 \times 113 = 17{,}345.6 \text{ N} = 17.35 \text{ kN}$$

ϕ6-mm stirrups at 200 mm o.c.

$$V_s = \frac{56.55 \times 450 \times 113}{200} = 14{,}377.5 \text{ N} = 14.38 \text{ kN}$$

$$V_n = V_c + V_s = 31.73 \text{ kN} < V_{exp} = 34.5 \text{ kN} \qquad \text{O.K.}$$

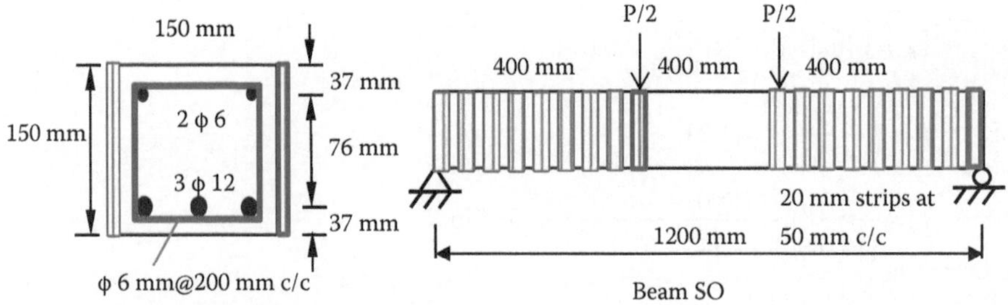

FIGURE 6.6 Example 6.4 showing the beam section and profile.

$$\Psi_f V_f = 0.85 \times \frac{A_{fv} f_{fe} d_{fv}}{s_f}$$

$$\varepsilon_{fu} = C_E \varepsilon_{fu}^* = 0.75 \times 0.01278 = 0.00959 \qquad \text{GFRP interior exposure}$$

$$k_1 = \left(\frac{37.7}{27}\right)^{2/3} = 1.25$$

$$L_e = \frac{23{,}300}{\left(1 \times 3 \times 15.65 \times 10^3\right)^{0.58}} = 45.48 \text{ mm}$$

$$k_2 = \frac{113 - 2 \times 45.48}{113} = 0.195$$

$$k_v = \frac{1.25 \times 0.195 \times 45.48}{11{,}900 \times 0.00959} = 0.097 < 0.75$$

$$\varepsilon_{fe} = k_v \varepsilon_{fu} = 9.30 \times 10^{-4} < 0.004$$

$$\Psi_f V_f = 0.85 \times \frac{2 \times 3 \times 20 \times 15{,}650 \times 0.000930 \times 113}{50} = 3{,}355.1 \text{ N} = 3.36 \text{ kN}$$

$$\bar{V}_n = V_c + V_s + \psi_f V_f = 31.73 \text{ kN} + 3.36 \text{ kN} = 35.09 \text{ kN} < \bar{V}_{exp} = 41.5 \text{ kN} \qquad \text{O.K.}$$

The ACI 440 model underestimates the shear capacity with FRP in this case, which is on the conservative side.

Example 6.5: Analysis

Khalifa and Nanni (2000) tested six reinforced concrete T-beams deficient in shear and strengthened in shear using CFRP sheets. The control beam was designated as BT1, and one of the strengthened beams was designated as BT4. This strengthened beam had no steel stirrups and had transverse CFRP U-wrap strips of 50-mm width at 125-mm on center extending the entire web height along the clear span. The geometric and material parameters of the beams are given in Figure 6.7.

$f'_c = 35$ MPa, $f_y = 470$ MPa for ϕ28-mm bars, and $f_y = 350$ MPa for ϕ13-mm and ϕ10-mm bars.
CFRP sheet: $f_{fu} = 3790$ MPa, $E_f = 228$ GPa, $t_f = 0.165$ mm

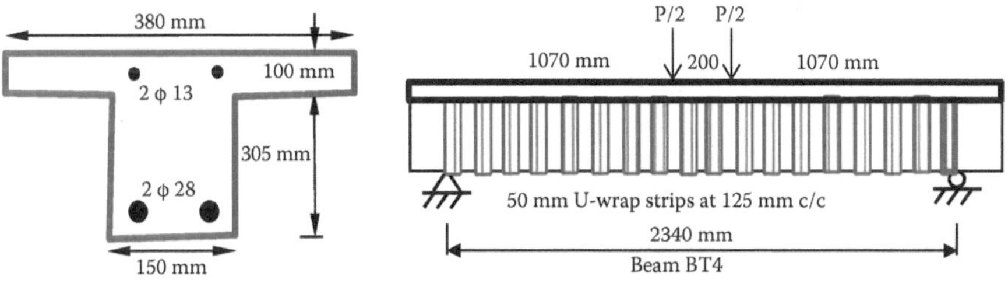

FIGURE 6.7 Example 6.5 showing the beam section and profile.

Solution:

The unstrengthened beams have no steel stirrups provided in the clear span.

$$d = 100 + 305 - 25 - 10 - \frac{28}{2} = 356 \text{ mm}$$

$$d_{fv} = 305 - 25 - 10 - \frac{28}{2} = 256 \text{ mm}$$

$$V_c = \frac{\sqrt{f_c'}}{6} b_w d = \frac{\sqrt{35}}{6} \times 150 \times 356 = 52{,}653.1 \text{ N} = 52.65 \text{ kN} < V_{exp} = 90 \text{ kN} \quad \text{O.K.}$$

$$\Psi_f V_f = 0.85 \times \frac{A_{fv} f_{fe} d_{fv}}{S_f}$$

$$\varepsilon_{fu} = C_E \varepsilon_{fu}^* = 0.95 \times \frac{3{,}790}{228{,}000} = 0.0158 \text{ CFRP interior exposure}$$

$$k_1 = \left(\frac{35}{27}\right)^{2/3} = 1.19$$

$$L_e = \frac{23{,}300}{\left(1 \times 0.165 \times 228 \times 10^3\right)^{0.58}} = 51.71 \text{ mm}$$

$$k_2 = \frac{256 - 51.71}{256} = 0.8$$

$$k_v = \frac{1.19 \times 0.8 \times 51.71}{11{,}900 \times 0.0158} = 0.262 < 0.75$$

$$\varepsilon_{fe} = k_v \varepsilon_{fu} = 0.262 \times 0.0158 = 0.00414 > 0.004$$

$$\Psi_f V_f = 0.85 \times \frac{2 \times 0.165 \times 50 \times 228{,}000 \times 0.004 \times 256}{125} = 26{,}195.6 \text{ N} = 26.2 \text{ kN}$$

$$\bar{V}_n = V_c + V_s + \psi_f V_f = 52.65 \text{ kN} + 0 + 26.2 \text{ kN} = 78.85 \text{ kN} < \bar{V}_{exp} = 162 \text{ kN} \quad \text{O.K.}$$

The ACI 440 model significantly underestimates the shear capacity with FRP in this case, which is considerably on the conservative side. However, it is important to note here that Beam BT5, which is identical to Beam BT4 except for using side strips, failed at a much lower capacity (121.5 kN). It is worthwhile to check the capacity of BT5 using the ACI440.2R-08 model (Problem 6.5 below) to see if it captures a similar drop in shear strength. The reader is referred to a relatively more recent analytical model using the truss analogy method for a more in-depth comparison with existing models (Colotti, Spadea, and Swamy 2004). The reader is also referred to a recent article assessing various design models for shear strengthening (Pellegrino and Vasic 2013).

Chapter Problems

Problem 6.1

The beam section in Problem 2.5 is deficient in shear, so strengthen the beam to resist enough shear such that it fails in flexure and not shear. The beam is under its own weight in addition to a single concentrated live load at mid-span, as shown in Figure 6.P.1.

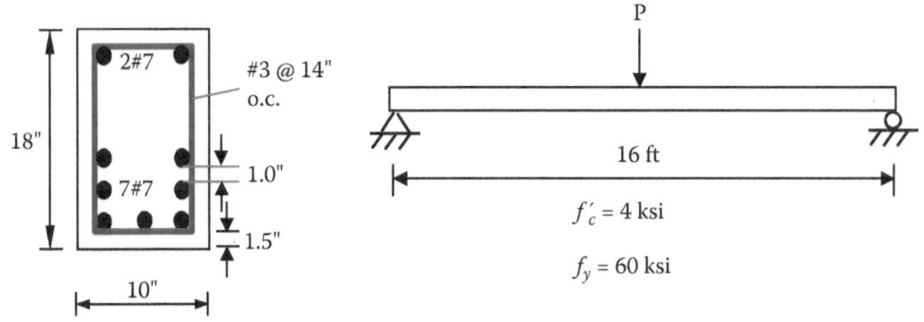

FIGURE 6.P.1

Use MBrace CF130 CFRP sheets Table 3.6.

Problem 6.2

Repeat Problem 6.1 using NSM CFRP tape (Aslan 500) with the properties shown in Figure 6.P.2. Note that the ACI440.2R-08 procedure for shear strengthening has been calibrated for FRP sheets and not NSM tape. However, follow the same procedure using NSM tape.

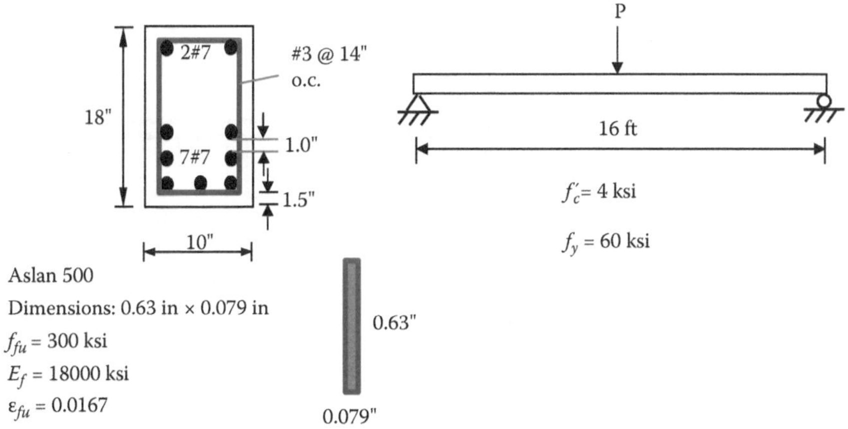

FIGURE 6.P.2

Problem 6.3

Assume that the beam given in Example 15.3 of the ACI440.2R-08 is deficient in shear (reinforced with #3 stirrups @ 12" c/c). Strengthen this beam in shear under the upgraded loads using the same FRP used in the example for flexural strengthening (Example 15.3 of ACI440.0R-08).

Problem 6.4

For the beam in Example 6.1, design the shear strengthening required if $\overline{w}_u = 7$ k/ft and the existing shear stirrups are #4 @ 10" c/c. See Figure 6.P.3.

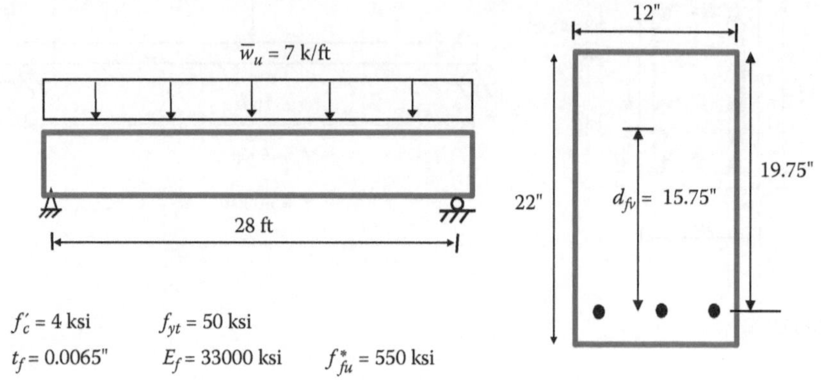

FIGURE 6.P.3

Problem 6.5

Khalifa and Nanni (2000) tested a deficient T-beam in shear (BT5) after strengthening it in shear using CFRP sheets. This strengthened beam had no steel stirrups and transverse CFRP side strips of 50-mm width at 125 mm on center extending the entire web height along the clear span. The geometric and material parameters of the beams are given in Figure 6.P.4.

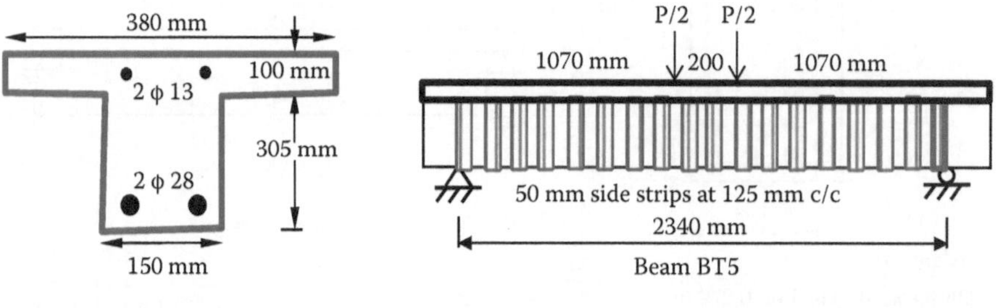

FIGURE 6.P.4

$f'_c = 35$ MPa, $f_y = 470$ MPa for ϕ28-mm bars, and $f_y = 350$ MPa for ϕ13-mm and ϕ10-mm bars

CFRP sheet: $f_{fu} = 3790$ MPa, $E_f = 228$ GPa, $t_f = 0.165$ mm

Check the ultimate shear capacity of BT5 and compare it with the actual ultimate shear strength of 121.5 kN.

Problem 6.6

Al-Sulaimani et al. (1994) tested a reinforced concrete beam (JO) deficient in shear and strengthened in shear using GFRP U-jackets covering more than the entire shear spans, as shown in Figure 6.P.5. The experiment indicates that this beam failed in flexure at 50.1 kN. Confirm that the shear capacity after strengthening is adequate to induce a flexural failure prior to failing in shear. The geometric and material parameters of the strengthened beam are given in Figure 6.P.5.

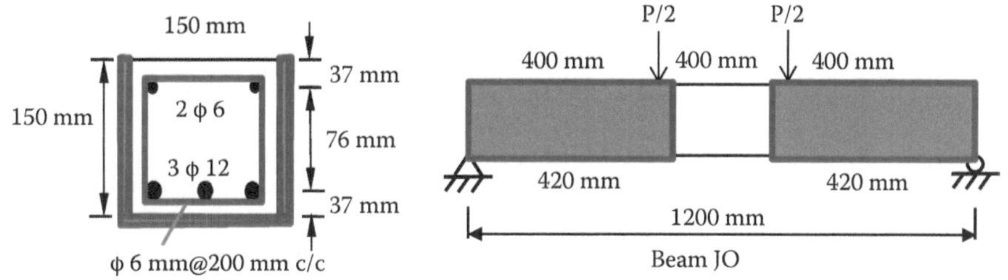

FIGURE 6.P.5

$f'_c = 37.7$ MPa, $f_y = 450$ MPa
GFRP plate: $f_{fu} = 200$ MPa, $E_f = 15.65$ GPa, $\varepsilon_{fu} = 0.01278$, $t_f = 3$ mm

Problem 6.7

Al-Sulaimani et al. (1994) tested a reinforced concrete beam (WO) deficient in shear and strengthened in shear using GFRP side plates covering more than the whole shear spans, as shown in Figure 6.P.6. The experiment indicates that this beam failed in shear at 42 kN. Compute the shear capacity after strengthening of this beam using ACI 440.2R-08 model. The geometric and material parameters of the strengthened beam are given in Figure 6.P.6.

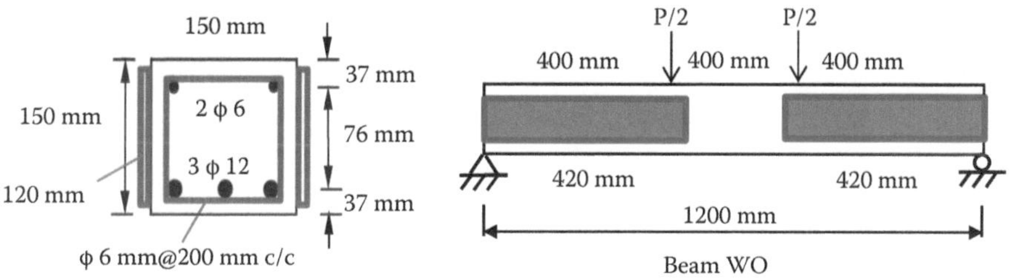

FIGURE 6.P.6

$f'_c = 37.7$ MPa, $f_y = 450$ MPa
GFRP plate: $f_{fu} = 200$ MPa, $E_f = 15.65$ GPa, $\varepsilon_{fu} = 0.01278$, $t_f = 3$ mm, $h_f = 120$ mm

REFERENCES

ACI 318-05. 2005. Building code requirements for structural concrete and commentary. ACI Committee 318, Farmington Hills, MI.

ACI 318-11. 2011. Building code requirements for structural concrete and commentary. ACI Committee 318, Farmington Hills, MI.

ACI 440.2R-08. 2008. Guide for the design and construction of externally bonded FRP systems for strengthening concrete structures. ACI Committee 440, Farmington Hills, MI.

Al-Sulaimani, G. J., A. Sharif, I. A. Basunbul, M. H. Baluch, and B. N. Ghaleb. 1994. Shear repair for reinforced concrete by fiberglass plate bonding. *ACI Structural Journal* 91 (3): 458–64.

Colotti, V., G. Spadea, and R. N. Swamy. 2004. Analytical model to evaluate failure behavior of plated reinforced concrete beams strengthened for shear. *ACI Structural Journal* 101 (6): 755–64.

Khalifa, A., W. Gold, A. Nanni, and M. Abel-Aziz. 1998. Contribution of externally bonded FRP to the shear capacity of RC flexural members. *ASCE Journal of Composites for Construction* 2 (4): 195–203.

Khalifa, A., and A. Nanni. 2000. Improving shear capacity of existing RC T-section beams using CFRP composites. *Cement and Concrete Composites* 22: 165–74.

Norris, T., H. Saadatmanesh, and M. R. Ehsani. 1997. Shear and flexural strengthening of R/C beams with carbon fiber sheets. *ASCE Journal of Structural Engineering* 123 (7): 903–11.

Pellegrino, C., and M. Vasic. 2013. Assessment of design procedures for the use of externally bonded FRP composites in shear strengthening of reinforced concrete beams. *Composites Part B: Engineering* 45:727–41.

Priestley, M., F. Seible, and G. Calvi. 1996. *Seismic design and retrofit of bridges*. New York: John Wiley and Sons.

Triantafillou, T. C. 1998. Shear strengthening of reinforced concrete beams using epoxy-bonded FRP composites. *ACI Structural Journal* 95 (2): 107–15.

Triantafillou, T. C., and C. P. Antonopoulos. 2000. Design of concrete flexural members strengthened in shear with FRP. *ASCE Journal of Composites for Construction* 4 (4): 198–205.

7 Strengthening of Columns for Confinement

7.1 OVERVIEW

FRP jackets are used to provide confinement to reinforced concrete columns, which improves both the strength and ductility. Confinement using FRP contributes to enhanced peak load resistance and enhanced rotation and drift ratio without a significant reduction in strength.

7.2 ENHANCEMENT OF PURE AXIAL COMPRESSION

The highest level of strength enhancement is obtained in the case of pure axial compression, as confinement activates a three-dimensional (3-D) state of stress under compression, which yields a higher axial capacity than that of unconfined compression.

FRP jackets are known to offer passive confinement to columns. This means that such confinement is not activated until dilation and cracking take place. Accordingly, intimate contact between the FRP wrapping and concrete column is important. This application is, therefore, called contact-critical application.

As the level of confinement increases, the uniaxial stress–strain behavior along the column axis changes from unconfined to lightly confined to a moderately confined softening curve to a heavily confined hardening curve (Figure 7.1).

Confinement of columns is accomplished by orienting the fibers transverse to the axis of the column such that it will act in a similar way as that of conventional spiral or tie steel reinforcement. The contribution of axially aligned fibers is beyond the scope of this chapter.

The peak strength in the axial direction (f'_{cc}) is the stress of the concrete section corresponding to the peak load after subtracting the contribution of the longitudinal steel reinforcement.

For columns with existing spiral steel reinforcement, ACI 440.2R-08 specifies the ultimate axial load as

$$\phi P_n = 0.85\phi\left[0.85 f'_{cc}\left(A_g - A_{st}\right) + A_{st} f_y\right] \tag{7.1}$$

For columns with existing tie steel reinforcement, ACI 440.2R-08 specifies the ultimate axial load as

$$\phi P_n = 0.8\phi\left[0.85 f'_{cc}\left(A_g - A_{st}\right) + A_{st} f_y\right] \tag{7.2}$$

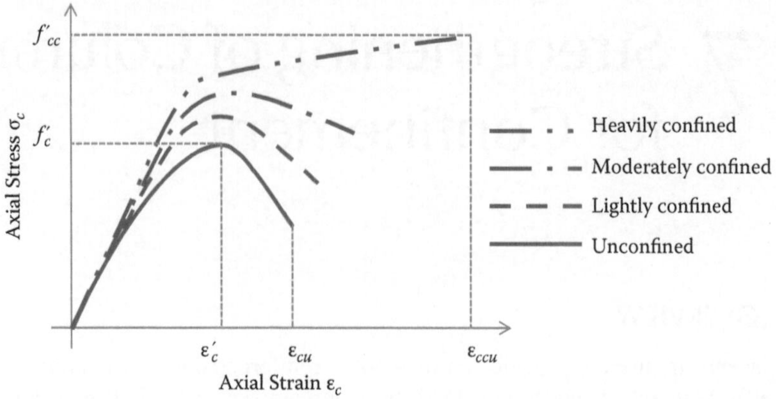

FIGURE 7.1 Stress–strain curves for unconfined and FRP-confined reinforced concrete columns.

where the strength-reduction factor $\phi = 0.75$ for spiral-reinforced members and $\phi = 0.65$ for tie-reinforced members (according to Figure 2.1) for compression-controlled sections, f'_{cc} is the maximum confined concrete strength due to FRP only (as per ACI 440.2R-08), A_g is the gross cross-sectional area of concrete, A_{st} is the area of longitudinal steel reinforcement in the column, and f_y is the yield strength of the longitudinal steel reinforcement.

ACI 440.2R-08 adopted the Lam and Teng (2003a) model, which takes into account FRP confinement only. Abd El Fattah (2012) modified the maximum confinement pressure (f_l) expression to account for both transverse steel and FRP inside the core (f_{le}) and only FRP in the cover (f_{lf}). The Lam and Teng (2003a) model is schematically presented in Figure 7.2.

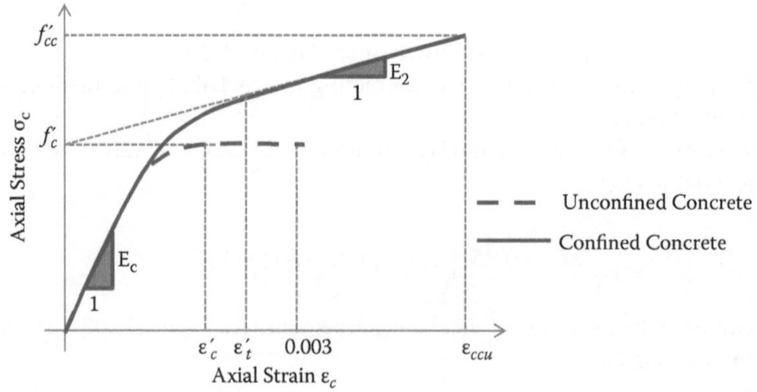

FIGURE 7.2 Lam and Teng (2003a) model for FRP-confined concrete under pure axial compression.

7.2.1 Lam and Teng Model

The equations that describe the Lam and Teng (2003a) model are given as follows:

$$f_c = \begin{cases} E_c\varepsilon_c - \dfrac{(E_c - E_2)^2}{4f_c'}\varepsilon_c^2 & 0 \le \varepsilon_c \le \varepsilon_t' \\ f_c' + E_2\varepsilon_c & \varepsilon_t' \le \varepsilon_c \le \varepsilon_{ccu} \end{cases} \tag{7.3}$$

$$E_2 = \frac{f_{cc}' - f_c'}{\varepsilon_{ccu}} \tag{7.4}$$

$$\varepsilon_t' = \frac{2f_c'}{E_c - E_2} \tag{7.5}$$

$$f_{cc}' = f_c' + \Psi_f \times 3.3 \times \kappa_a f_l \tag{7.6}$$

$$f_l = \frac{2nt_f E_f \varepsilon_{fe}}{D} \tag{7.7}$$

where $\Psi_f = 0.95$ is an additional reduction factor added by ACI 440.2R-08, κ_a is a strength efficiency factor to account for the section geometry ($\kappa_a = 1.0$ for a circular section, while its computation for a noncircular section is discussed later in this chapter), and $\varepsilon_{fe} = \kappa_\varepsilon \varepsilon_{fu} = 0.586\varepsilon_{fu}$ as averaged by Lam and Teng (2003a). Others confirmed this reduction by obtaining κ_ε from experiments in the range of 0.57–0.61 (Carey and Harries 2005). Pessiki et al. (2001) attributed this strain reduction to the multiaxial state of stress that the FRP is subjected to in this application. *D* is the diameter of the circular section or the diagonal of the noncircular section, as discussed in Section 7.2.2. According to Lam and Teng (2003a,b), the minimum confinement ratio (f_l/f_c') should exceed 0.07 in order for the confined axial stress–strain diagram of circular columns to have an ascending second branch, as seen in Figure 7.2. For noncircular sections, the ratio (f_l/f_c') is multiplied by (κ_a) with the product to exceed 0.07 in order to have an ascending second branch. On the other hand, ACI 440.2R-08 increased this minimum limit to 0.08 to guarantee the outcome of an ascending curve only while equally applying (f_l/f_c') to circular and noncircular sections. Abd El Fattah (2012) allowed this ratio (f_l/f_c') to drop below 0.08, in which case a descending branch of the stress–strain curve is accounted for by using a Mander constitutive model (Mander et al. 1988). The ultimate axial column strain is found by the empirical formula suggested by ACI 440.2R-08.

$$\varepsilon_{ccu} = \varepsilon_c'\left(1.50 + 12\kappa_b \frac{f_l}{f_c'}\left(\frac{\varepsilon_{fe}}{\varepsilon_c'}\right)^{0.45}\right) \tag{7.8}$$

$$\varepsilon_{ccu} \le 0.01 \tag{7.9}$$

where κ_b is a strain efficiency factor to account for the section geometry ($\kappa_b = 1.0$ for a circular section, while its computation for a noncircular section is discussed later). The maximum axial compressive strain (ε_{ccu}) is limited by ACI 440.2R-08 to a value of 0.01, as shown in Equation (7.9), to prevent excessive cracking and possible loss of concrete integrity. ACI 440.2R-08 also limits this strength improvement technique to concrete having f'_c less than 10 ksi (69 MPa) due to the lack of experimental studies with concrete of higher values. It is important to state here that in the case where ε_{ccu} is controlled by Equation (7.9), the slope of the second hardening line (E_2) in Equation (7.4) will be kept constant, and the value of (f'_{cc}) will be reduced accordingly (ACI 440.2R-08).

7.2.2 Consideration of Rectangular Sections

ACI 440.2R-08 confirmed the findings of several researchers that confining square and rectangular columns with FRP jackets may provide a marginal increase in the maximum confined axial compressive strength (f'_{cc}). The provisions of ACI 440.2R-08 ignore the contribution of conventional steel transverse reinforcement in confining the columns, and they limit this application to columns having *h*/*b* aspect ratio less than or equal to 2.0, with both *h* and *b* less than 36 in. (914 mm). ACI 440.2R-08 adopts the theoretical model proposed by Lam and Teng (2003b) that accounts for the reduced area of the confined rectangular column. The rectangular section is converted into an equivalent circular section for the purpose of calculating the maximum confining pressure (f_l) using Equation (7.7), with the diameter *D* replaced to be the diagonal of the rectangular section (Figure 7.3).

$$D = \sqrt{b^2 + h^2} \tag{7.10}$$

To determine the area of the cross section that is effectively confined by the FRP jacket, four parabolas are drawn inside the rectangular section to isolate the inner confined area from the outer unconfined area. The parameters controlling the sizes of the parabolas are the column dimensions (*b*, *h*); the radius of the corners of the

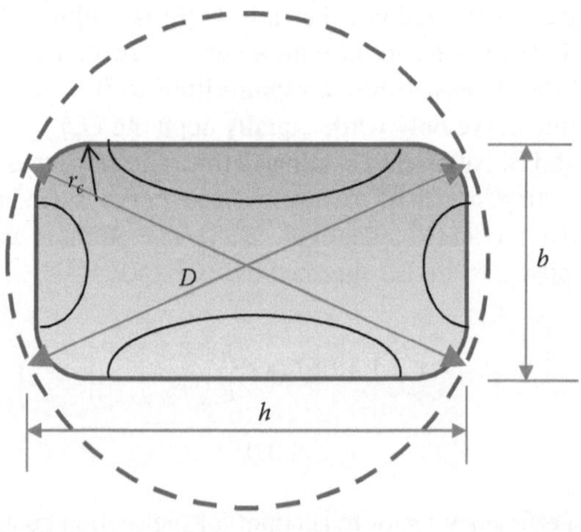

FIGURE 7.3 Effective area of confined rectangular section per ACI 440.2R-08. (Courtesy of Lam and Teng [2003b].)

column (r_c), which is necessary to introduce to the column to provide some meaningful confinement effect; and the longitudinal steel reinforcement ratio (ρ_s). It can be easily shown that the effective area to concrete area ratio (A_e/A_c) is found to be

$$\frac{A_e}{A_c} = \frac{1 - \dfrac{\left[\left(\dfrac{b}{h}\right)(h-2r_c)^2 + \left(\dfrac{h}{b}\right)(b-2r_c)^2\right]}{3A_g} - \rho_s}{1-\rho_s} \tag{7.11}$$

The strength and strain efficiency factors (κ_a, κ_b) are functions of the effective area of confined rectangular section (A_e), determined from Equation (7.11), and the column aspect ratio (b/h) as postulated by Lam and Teng (2003b). The respective expressions of these variables are given by Equations (7.12) and (7.13):

$$\kappa_a = \frac{A_e}{A_c}\left(\frac{b}{h}\right)^2 \tag{7.12}$$

$$\kappa_b = \frac{A_e}{A_c}\left(\frac{h}{b}\right)^{0.5} \tag{7.13}$$

7.2.3 Combined Confinement of FRP and Transverse Steel in Circular Sections

The f_l expression given by Equation (7.7) is updated by Abd El Fattah (2012) to account for the contribution of FRP and transverse steel in confining the core:

$$f_{le} = \frac{2nt_f E_f \varepsilon_{fe}}{D} + \frac{2k_e A_{sp} f_{yh}}{d_c \times S} \tag{7.14}$$

where $k_e = \dfrac{1-\frac{s'}{2d_c}}{1-\rho_{cc}}$ for spiral transverse steel and $k_e = \dfrac{\left(1-\frac{s'}{2d_c}\right)^2}{1-\rho_{cc}}$ for circular hoop transverse steel, s' is the clear spacing of hoops or clear pitch of spiral, d_c is the diameter of concrete core c/c, and ρ_{cc} is the ratio of longitudinal reinforcement area to the area of core.

Similarly, the f_l expression that confines the cover of the column only is determined by

$$f_{lf} = \frac{2nt_f E_f \varepsilon_{fe}}{D} \tag{7.15}$$

The values of the ultimate confined strength of the core and cover (f'_{cce} and f'_{ccf}) are determined from Equation (7.6) by substituting f_{le} and f_{lf}, respectively.

Accordingly, the ultimate confined concrete axial load, replacing that of Equation (7.1), is

$$\phi P_n = 0.85\phi\left[0.85 f'_{cce}(A_c - A_{st}) + 0.85 f'_{ccf}(A_g - A_c) + A_{st} f_y\right] \quad (7.16)$$

For the combined confinement effect when f_{lf}/f'_c is greater than 0.08, the Lam and Teng (2003a) model described in Section 7.2.1 is used with the proper f'_{cce} and f'_{ccf} for the core and the cover, respectively. In this case, the ultimate compressive strain (ε_{ccu}) is assumed to be different in the core and cover, which is determined by substituting f_{lf} or f_{le} for f_l in Equation (7.8). Accordingly, the slope (E_{2e}) of the core will also be different from the slope of the cover (E_{2f}). On the other hand, when f_{lf}/f'_c is less than 0.08, the Mander model described in Mander, Priestley, and Park (1988) is used with the proper f'_{cce} and f'_{ccf} for the core and the cover, respectively. The minimum ultimate compressive strain (ε_{ccu})—between that of Equations (7.8) and (7.9) using f_{lf} for the case of FRP and that of the energy approach corresponding to the fracture of the first hoop (Mander et al. 1988) for the case of transverse steel—is selected. So in this case, the peaks of the curves for the core and cover take place at different strains (ε'_{cce} and ε'_{ccf}) corresponding to the different strength values for the core and the cover (f'_{cce} and f'_{ccf}), and the ultimate strain (ε_{ccu}) is also different for both the core and the cover. This procedure has been programmed into the software "KDOT Column Expert" developed by the author and coworkers and described in two references (Abd El Fattah 2012; Rasheed et al. 2012).

7.2.4 Combined Confinement of FRP and Transverse Steel in Rectangular Sections

The f_l expression given by Equation (7.7) is replaced with the two lateral pressures along the x- and y-axes as formulated by Al-Rahmani and Rasheed (2014) to account for the contribution of FRP and transverse steel in confining the core and the cover in the x- and y-directions, as seen in Figure 7.4.

$$f_{lxe} = k'_e \frac{2nt_f E_f \varepsilon_{fe}}{h} + k_e \rho_x f_{yh} \quad (7.17)$$

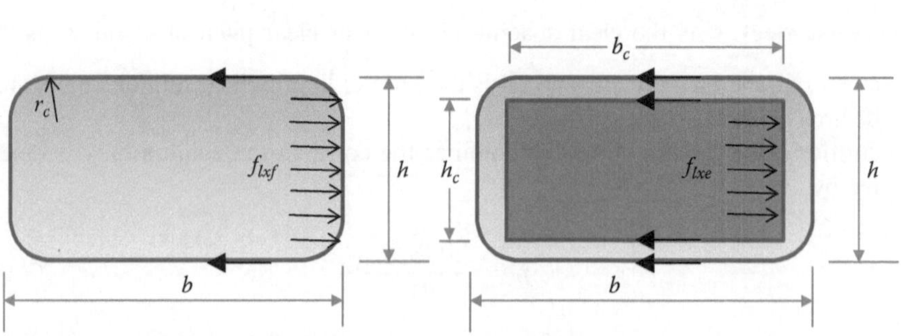

FIGURE 7.4 Effective horizontal pressure of confined rectangular section. (Courtesy of Al-Rahmani and Rasheed [2014].)

$$f_{lye} = k'_e \frac{2nt_f E_f \varepsilon_{fe}}{b} + k_e \rho_y f_{yh} \tag{7.18}$$

$$f_{lxf} = k'_e \frac{2nt_f E_f \varepsilon_{fe}}{h} \tag{7.19}$$

$$f_{lyf} = k'_e \frac{2nt_f E_f \varepsilon_{fe}}{b} \tag{7.20}$$

where $k'_e = \frac{A_e}{A_c}$ from Equation (7.11), $k_e = \frac{\left(1 - \sum_{i=1}^{n} \frac{(w'_i)^2}{6b_c d_c}\right)\left(1 - \frac{s'}{2b_c}\right)\left(1 - \frac{s'}{2d_c}\right)}{(1 - \rho_{cc})}$ for rectangular tied transverse steel, $\rho_x = \frac{2A_{tie}}{h_c \times s}, \rho_y = \frac{2A_{tie}}{b_c \times s}$ for rectangular tie transverse steel, s is the c/c spacing of ties, h_c and b_c are the height and width of concrete core c/c, and ρ_{cc} is the ratio of longitudinal reinforcement area to the area of core (Mander et al. 1988).

The values of the ultimate confined strength of the core and cover (f'_{cce}, and f'_{ccf}) are determined based on the 3-D state of stress of concrete plasticity proposed by Willam and Warnke (1975) and explained below by substituting f_{lxe} and f_{lye} as lateral core pressures and (f_{lxf} and f_{lyf}) as lateral cover pressures, respectively. Accordingly, the ultimate confined concrete axial load replacing that of Equation (7.2) is

$$\phi P_n = 0.8\phi\left[0.85 f'_{cce}(A_c - A_{st}) + 0.85 f'_{ccf}(A_g - A_c) + A_{st} f_y\right] \tag{7.21}$$

In selecting whether to use Lam and Teng's ascending model or Mander's descending model, f_{lf} of the equivalent circular column described in Section 7.2.2 and substituted into Equation (7.7) will be computed. If f_{lf}/f'_c is greater than 0.08, Lam and Teng's ascending model described in Section 7.2.1 is used, with the proper f'_{cce}, and f'_{ccf} for the core and the cover determined using the 3-D state-of-stress concrete plasticity model (Willam and Warnke 1975) explained in Section 7.2.5. In this case, the ultimate compressive strain (ε_{ccu}) is assumed to be different in the core and cover, determined by substituting f_{le} and f_{lf} for f_l in Equation (7.8), respectively. Accordingly, the slope (E_{2e}) of the core will be higher than the slope of the cover (E_{2f}).

On the other hand, when f_{lf}/f'_c is less than 0.08, Mander's model (Mander et al. 1988) is used with the proper (f'_{cce}, and f'_{ccf}) for the core and the cover determined using the 3-D state-of-stress concrete plasticity model (Willam and Warnke 1975) explained in section 7.25. The ultimate compressive strains ε_{ccue} and ε_{ccuf} for the core and the cover are selected. ε_{ccuf} is taken from Equations (7.8) and (7.9) using f_{lf} for the case of the cover and ε_{ccue} is taken from the energy approach corresponding to the fracture of the first hoop (Mander et al. 1988) for the case of the core. The peak of the curve also takes place at different strains ε'_{cce}, and ε'_{ccf} corresponding to f'_{cce}, and f'_{ccf}, with different strength values for the core and the covers. This procedure has been programmed into the software "KDOT Column Expert" developed by the author and coworkers and described by Al-Rahmani and Rasheed (2014).

7.2.5 3-D State of Stress Concrete Plasticity Model

Mander (1983) proposed using a multiaxial stress procedure to calculate the ultimate confined strength (e.g., f'_{cce}) from two given lateral pressures (e.g., f_{lxe} and f_{lye}). This numerical procedure is summarized in the following steps:

1. Determine f_{lxe} and f_{lye} using Equations (7.17) and (7.18)
2. Convert the positive sign of f_{lxe} and f_{lye} to negative to represent the major and intermediate principal stresses (these values are referred to as σ_1 and σ_2 so that $\sigma_1 > \sigma_2$)
3. Estimate the confined strength f'_{cce}, which is σ_3 as the minor principal stress
4. Calculate the octahedral stress σ_{oct}, octahedral shear stress τ_{oct}, and lode angle θ as follows:

$$\sigma_{oct} = \frac{1}{3}(\sigma_1 + \sigma_2 + \sigma_3) \tag{7.22}$$

$$\tau_{oct} = \frac{1}{3}\left[(\sigma_1 - \sigma_2)^2 + (\sigma_2 - \sigma_3)^2 + (\sigma_3 - \sigma_1)^2\right]^{\frac{1}{2}} \tag{7.23}$$

$$\cos\theta = \left[\frac{\sigma_1 - \sigma_{oct}}{\sqrt{2}\tau_{oct}}\right] \tag{7.24}$$

5. Determining the ultimate strength meridian surfaces T and C (for $\theta = 0°$ and $60°$, respectively) using the following equations derived by Al-Rahmani and Rasheed (2014) from data by Kupfer, Hilsdorf and Rüsch (1969) while calibrating the data against the equivalent circular section of Lam and Teng (2003b):

$$T = \begin{cases} 0.061898 - 0.62637\bar{\sigma}_{oct} & \text{if } \bar{\sigma}_{oct} > -0.767 \\ 0.229132 - 0.40824\bar{\sigma}_{oct} & \text{if } \bar{\sigma}_{oct} \le -0.767 \end{cases} \tag{7.25}$$

$$C = \begin{cases} 0.107795 - 1.09083\bar{\sigma}_{oct} & \text{if } \bar{\sigma}_{oct} > -0.333 \\ 0.336883 - 0.40357\bar{\sigma}_{oct} & \text{if } \bar{\sigma}_{oct} \le -0.333 \end{cases} \tag{7.26}$$

$$\bar{\sigma}_{oct} = \sigma_{oct}/f'_c \tag{7.27}$$

6. Determining the octahedral shear stress using the interpolation function found by Willam and Warnke (1975):

$$\bar{\tau}_{oct} = C\frac{0.5D/\cos\theta + (2T - C)\left[D + 5T^2 - 4TC\right]^{\frac{1}{2}}}{D + (2T - C)^2} \tag{7.28}$$

$$D = 4\left(C^2 - T^2\right)\cos^2\theta \tag{7.29}$$

$$\tau_{oct} = \bar{\tau}_{oct} f_c' \tag{7.30}$$

7. Recalculating (f'_{cce}) using the following equation (same as Equation [7.23], but solving for σ_3):

$$\sigma_3 = \frac{\sigma_1 + \sigma_2}{2} - \sqrt{4.5\tau_{oct}^2 - 0.75(\sigma_1 - \sigma_2)^2} \tag{7.31}$$

8. If the value from Equation (7.31) matches that of the assumed initial value, then convergence is achieved. Otherwise, the value from Equation (7.31) is set as the initial value and steps 4 through 8 are repeated until convergence is accomplished.

An Excel spreadsheet program automating this procedure may be obtained electronically from the author.

7.3 ENHANCEMENT UNDER COMBINED AXIAL COMPRESSION AND BENDING MOMENT

It has been shown by experiments (Chaallal and Shahawy 2000; Memon and Sheikh 2005; Darby et al. 2011) and by computations (ACI 440.2R-08) that the use of FRP wrapping of circular, square, and slightly rectangular columns provides strength improvement under the effect of axial compression and bending moment. To establish a procedure accounting for FRP confinement effects on the strength improvement in wrapped columns, ACI 440.2R-08 takes the following considerations in constructing the P-M interaction diagram:

1. The equations of Section 7.2.1 and 7.2.2 are applicable to finding the pure compression point (Point A), see Figure 7.5.
2. The effective hoop FRP strain in the jacket for diagram points other than pure axial compression needs to assume a value of 0.004 in the compression-controlled region of the curve to ensure shear integrity of confined concrete. The ACI 440.2R-08 guide allows finding two more points to draw the compression-controlled region when establishing the interaction diagram, as seen in Figure 7.5:
   a. Point B with a strain distribution corresponding to a compressive strain of ε'_{ccu} at the extreme compression fiber and zero strain at the extreme layer of tensile steel reinforcement
   b. Point C with a strain distribution corresponding to a compressive strain of ε'_{ccu} at the extreme compression fiber and yielding tensile strain ε_{sy} at the extreme layer of tensile steel reinforcement (balanced failure)
   c. Connecting Points A, B, and C with straight lines to construct the enhanced compression-controlled region.

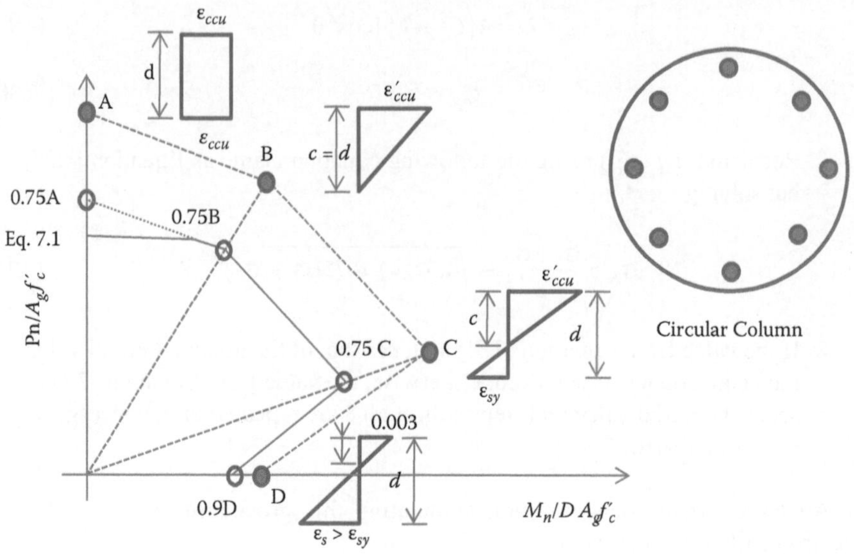

FIGURE 7.5 Simplified interaction diagrams for FRP-confined concrete circular column.

   d. Connecting Point C and the pure bending point with a straight line indicating no strength enhancement in the tension-controlled region.
3. Drawing the interaction diagram reduced by the relevant resistance (ϕ) factors per ACI 318-11 (2011) for circular and rectangular cross sections.
4. Using Equations (7.1) and (7.2) to cap the reduced interaction diagram when the eccentricity is less than or equal to 0.1 h.

7.3.1 Interaction Diagrams for Circular Columns

Since the derivation of the force and moment equations for points B and C requires the integration of some involved integrands with trigonometric functions due to the continuous variation of the section width, these expressions will be evaluated numerically by dividing the cross section into 100 layers parallel to the centroidal x-axis. The addition of the longitudinal steel bar contribution will have to be accounted for numerically anyway. Accordingly, the formulation of this procedure is incorporated into an Excel spreadsheet as follows.

7.3.1.1 Contribution of Concrete

The following computation steps are taken in general:

1. The thickness of each layer is

$$t_{lr} = \frac{D}{100} \tag{7.32}$$

2. The depth of each layer measured from the top extreme fiber is

$$d_{lr1} = \frac{t_{lr}}{2}, \quad d_{lr2} = \frac{t_{lr}}{2} + t_{lr}, \quad d_{lri} = \frac{2i-1}{2} \times t_{lr} \tag{7.33}$$

3. The angle of each layer measured from the vertical axis is

$$\cos\theta_{lri} = \frac{D/2 - d_{lri}}{D/2} \tag{7.34}$$

4. The width of each layer (i) is

$$b_{lri} = D \times \sin\theta_{lri} \tag{7.35}$$

5. The depth of the neutral axis (c) measured from the top extreme fiber is

$$c = d \text{ for point B} \tag{7.36}$$

$$c = d\frac{\varepsilon'_{ccu}}{\varepsilon'_{ccu} + \varepsilon_{sy}} \text{ for point C} \tag{7.37}$$

6. The depth of the transition point (d_t) measured from the top extreme fiber is

$$d_t = c - y_t = c\left(1 - \frac{\varepsilon'_t}{\varepsilon'_{ccu}}\right) \tag{7.38}$$

   where ε'_t is computed from Equation (7.5).
7. If $d_{lri} < d_t$, use Equation (7.3b); otherwise, use Equation (7.3a), with

$$\varepsilon_{ci} = \frac{c - d_{lri}}{c}\varepsilon'_{ccu} \tag{7.39}$$

   and f_{ci} from Equation (7.3).
8. The axial force (P_{cn}) and the bending moment (M_{cn}) contribution of concrete are

$$P_{cn} = \sum_i f_{ci} b_{lri} t_{lr}$$

$$M_{cn} = \sum_i f_{ci} b_{lri} t_{lr} \times \left(\frac{D}{2} - d_{lri}\right) \tag{7.40}$$

It is important to mention here that f'_{cc} and ε'_{ccu} for points B and C are determined from Equations (7.6), (7.8), and (7.9) by computing f_l in Equation (7.7) using $\varepsilon_{fe} = 0.004$ instead of $\kappa_\varepsilon \varepsilon_{fu}$ used for pure axial compression.

7.3.1.2 Contribution of Steel

The following computation steps are taken in general:

1. The angle between each two consecutive bars is

$$\theta_{bar} = \frac{360}{\text{No. of bars}} \tag{7.41}$$

2. The radius of bars is

$$R_{bar} = \frac{D}{2} - \text{clear cover} - d_{spiral} - \frac{d_{bar}}{2} \tag{7.42}$$

3. The depth of each bar measured from the top extreme fiber is

$$d_{\_barj} = \frac{D}{2} + R_{bar} \cos(j-1) \times \theta_{bar} \tag{7.43}$$

Note that the first bar is placed on the vertical axis on the tension side of the section (bottom side).

4. The strain and stress in the steel bar is

$$\varepsilon_{sj} = \frac{c - d_{\_barj}}{c} \varepsilon'_{ccu}$$
$$f_{sj} = E_s \varepsilon_{sj} \le f_y \tag{7.44}$$

and f_{cj} from Equation (7.3) by substituting ε_{sj} for ε_c.

5. The axial force (P_{sn}) and the bending moment (M_{sn}) contribution of steel are

$$P_{sn} = \sum_j (f_{sj} - f_{cj}) \times A_{s,bar}$$
$$M_{sn} = \sum_j (f_{sj} - f_{cj}) \times A_{s,bar} \times \left(\frac{D}{2} - d_{\_barj}\right) \tag{7.45}$$

The axial force and bending moment capacity of the section is the simple sum of the contribution of concrete and steel:

$$P_n = P_{cn} + P_{sn}$$
$$M_n = M_{cn} + M_{sn} \tag{7.46}$$

7.3.2 Interaction Diagrams for Circular Columns Using KDOT Column Expert

As shown in the previous section, the effective confining pressure (f_l) in the case of the two eccentric points (Point B and Point C) is lower than that of pure axial compression (Point A), where the section is fully confined. In KDOT Column Expert software, this issue of partial confinement is modeled more consistently throughout the range of eccentricities. While the case of pure axial compression has zero eccentricity and full confinement, the pure bending case has infinite eccentricity and no confinement at all. The confined strength in between the two extremes (f'_{cc} and f'_c) is mapped gradually as a function of the eccentricity:

$$\overline{f_{cc}} = \frac{1}{1+\frac{e}{D}} f'_{cc} + \frac{1}{1+\frac{D}{e}} f'_c \tag{7.47}$$

where $\overline{f_{cc}}$ is the eccentric confined strength at eccentricity (e/D), and the equation satisfies the two extremes (f'_{cc} and f'_c). Figure 7.6 illustrates three different sections under concentric load, a combination of axial load and bending moment, and pure bending moment: The highlighted fiber in the three cases has the same strain. However, the size of the compression zone does play an important role in predicting the stress, which is different in the three cases of Figure 7.6. Hence, it is more realistic to relate the eccentric strength and ductility to the level of confinement utilization and compression zone size represented in circular columns by the eccentricity, as seen in Figure 7.7.

7.3.2.1 Eccentric Model Based on Lam and Teng Equations

The ultimate eccentric or partially confined strength $\overline{f_{cc}}$ is determined from Equation (7.47) and is paired with the ultimate eccentric or partially confined strain

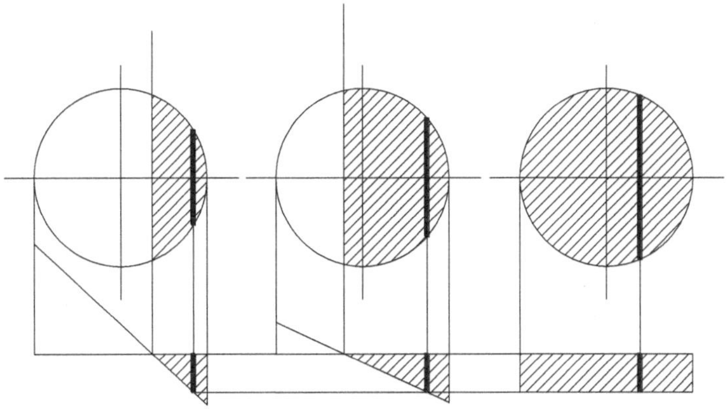

FIGURE 7.6 Effect of compression zone size or eccentricity on concrete strength.

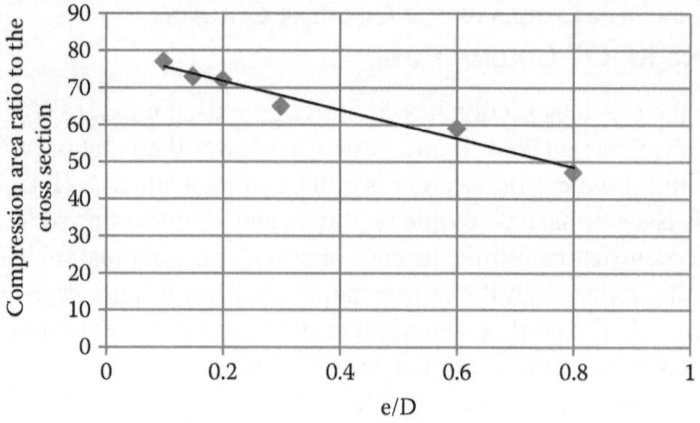

FIGURE 7.7 The relationship between the normalized compression zone size and eccentricity in circular columns. (Courtesy of Abd El Fattah [2012].)

$\overline{\varepsilon_{cu}}$ to construct a Lam and Teng eccentric curve. The strain $\overline{\varepsilon_{cu}}$ is found by linear interpolation between the two extreme bounds of strain, as seen in Figure 7.8:

$$\overline{\varepsilon}_{cu} = \frac{\overline{f}_{cc} - f_c'}{f_{cc}' - f_c'}(\varepsilon_{ccu} - 0.003) + 0.003 \tag{7.48}$$

Any point on the generated eccentric curves can be calculated using the following equations:

$$f_c = E_c\varepsilon_c - \frac{\left(E_c - \overline{E}_2\right)^2}{4f_c'}\varepsilon_c^2 \quad \text{for} \quad 0 \le \varepsilon_c \le \overline{\varepsilon}_t' \tag{7.49}$$

$$f_c = f_c' + \overline{E}_2\varepsilon_c \quad \text{for} \quad \overline{\varepsilon}_t' \le \varepsilon_c \le \overline{\varepsilon}_{cu} \tag{7.50}$$

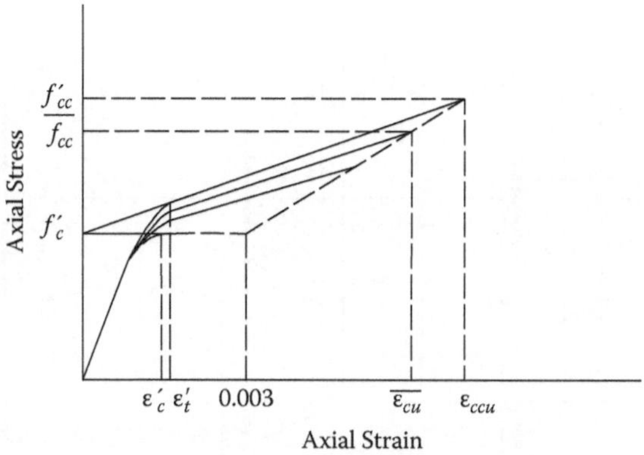

FIGURE 7.8 Eccentricity-based confined Lam and Teng model.

$$\bar{E}_2 = \frac{\bar{f}_{cc} - f'_c}{\bar{\varepsilon}_{cu}} \tag{7.51}$$

$$\bar{\varepsilon}'_t = \frac{2f'_c}{E_c - \bar{E}_2} \tag{7.52}$$

7.3.2.2 Eccentric Model Based on Mander Equations

The ultimate eccentric or partially confined strength $\overline{f_{cc}}$ is determined from Equation (7.47) and is paired with the corresponding strain $\overline{\varepsilon_{cc}}$ to construct a Mander eccentric curve. The corresponding strain $\overline{\varepsilon_{cc}}$ is given by

$$\bar{\varepsilon}_{cc} = \varepsilon_{co}\left[1 + 5\left(\frac{\bar{f}_{cc}}{f'_c} - 1\right)\right] \tag{7.53}$$

and the ultimate strain $\overline{\varepsilon_{cu}}$ corresponding to a specific eccentricity will be determined from a linear function between the ultimate point of the fully confined concrete (f_{cu}, ε_{cu}) and the ultimate point of unconfined concrete f_{cu0}, $\varepsilon_{cu0} = 0.003$, as seen in Figure 7.9.

$$\bar{\varepsilon}_{cu} = \bar{\varepsilon}_{cc}\left[\frac{\frac{\bar{E}_{sec}}{E_{sec,u}}\bar{r}}{\frac{c}{\bar{\varepsilon}_{cu}} + 1} - \bar{r} + 1\right]^{\frac{1}{\bar{r}}} \qquad E_{sec,u} = \frac{f_{cu} - f_{cu0}}{\varepsilon_{cu} - 0.003} \tag{7.54}$$

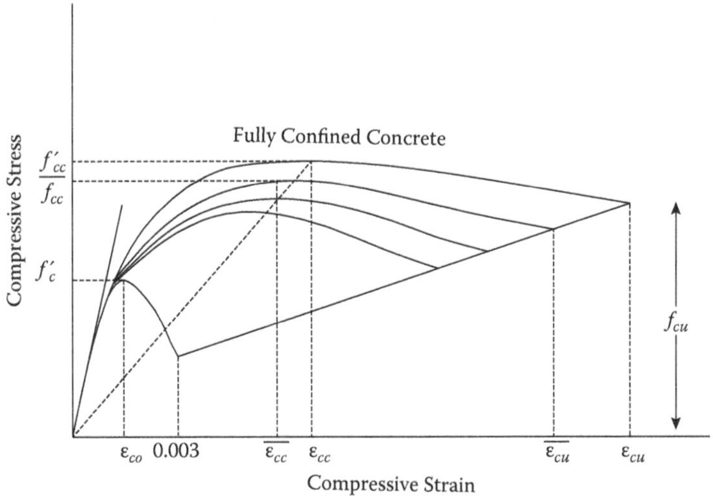

FIGURE 7.9 Eccentricity-based confined Mander model.

$$c = \frac{f_{cu} - E_{\sec,u} \times 0.003}{E_{\sec,u}} \tag{7.55}$$

$$\bar{E}_{\sec} = \frac{\bar{f}_{cc}}{\bar{\varepsilon}_{cc}} \quad \text{and} \quad \bar{r} = \frac{E_c}{E_c - \bar{E}_{\sec}} \tag{7.56}$$

Any point on the generated eccentric curve can be calculated using the following equation:

$$f_c = \frac{\bar{f}_{cc}\,\bar{x}\,\bar{r}}{\bar{r} - 1 + \bar{x}^{\bar{r}}} \tag{7.57}$$

where

$$\bar{x} = \frac{\varepsilon_c}{\bar{\varepsilon}_{cc}} \tag{7.58}$$

$$\bar{r} = \frac{E_c}{E_c - \bar{E}_{\sec}} \tag{7.59}$$

$$\bar{E}_{\sec} = \frac{\bar{f}_{cc}}{\bar{\varepsilon}_{cc}} \tag{7.60}$$

7.3.2.3 Eccentric-Based Model Selection

Eccentric-based model selection depends on the ratio of (f_{lf}/f'_c) for a concentrically loaded column. If this ratio (f_{lf}/f'_c) > 0.08, the Lam and Teng eccentric model is used in the analysis. If the ratio (f_{lf}/f'_c) < 0.08, the Mander eccentric model is used in the analysis, as seen in Figure 7.10. Accordingly, KDOT Column Expert allows the case of a descending curve to be analyzed.

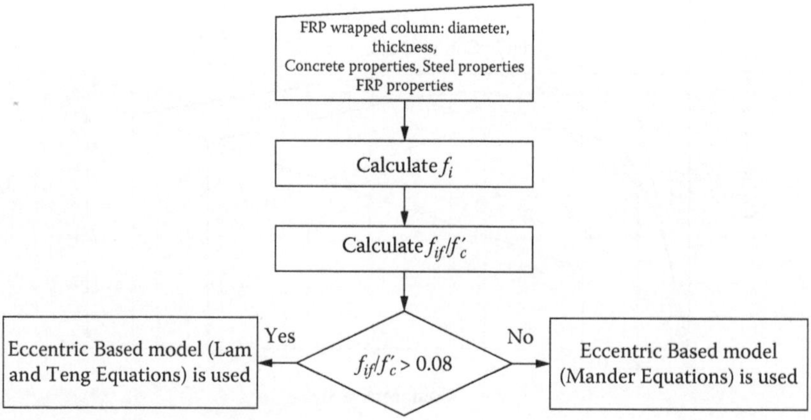

FIGURE 7.10 Eccentric model implementation strategy. (Courtesy of Abd El Fattah [2012].)

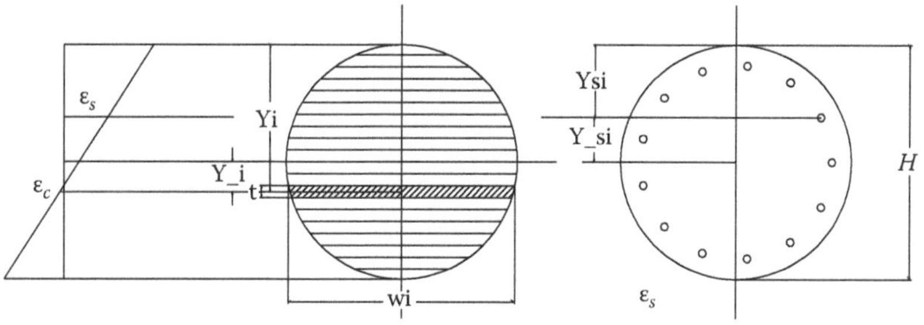

FIGURE 7.11 Using finite-layer approach in analysis.

7.3.2.4 Numerical Procedure

The column cross section is divided into a finite number of thin layers, as seen in Figure 7.11. The force and moment of each layer is calculated and stored. The bars are treated as discrete objects in their actual locations. The advantage of that is to precisely calculate the internal forces induced by steel bars and concrete layers in the column cross section. The cross section analyzed is loaded incrementally by maintaining a certain eccentricity between the axial force P and the resultant moment M_R. Since increasing the load and resultant moment causes the neutral axis and centroid to vary nonlinearly, the generalized moment-of-area theorem is devised. The method is developed using an incremental iterative analysis algorithm, a secant stiffness approach, and proportional or radial loading. It is explained in the following steps:

1. Calculate the initial section properties.
   Elastic axial rigidity, EA:

$$EA = \sum_i E_c w_i t_i + \sum_i (E_s - E_c) A_{si} \tag{7.61}$$

where E_c = initial modulus of elasticity of the concrete and E_s = initial modulus of elasticity of the steel bar.

The depth of the elastic centroid position from the bottom fiber of the section Y_c:

$$Y_c = \frac{\sum_i E_c w_i t_i (H - Y_i) + \sum_i (E_s - E_c) A_{si} (H - Y_{si})}{EA} \tag{7.62}$$

Elastic flexural rigidity about the elastic centroid, EI:

$$EI = \sum_i E_c w_i t_i (H - Y_i - Y_c)^2 + \sum_i (E_s - E_c) A_{si} (H - Y_{si} - Y_c)^2 \tag{7.63}$$

Typically, $Y_c = Y_G = \dfrac{H}{2}$

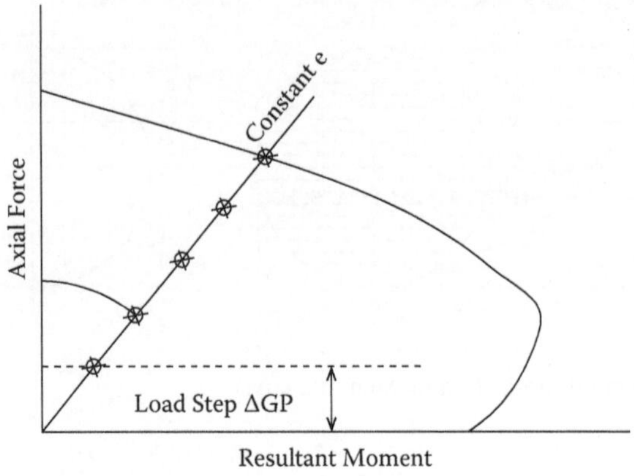

FIGURE 7.12 Radial loading concept.

2. Calculate f_{lf}/f_c' and check the ratio to decide which model is used (eccentric model based on Lam and Teng equations or eccentric model based on Mander equations), as seen in Figure 7.10.
3. Define eccentricity e, which specifies the radial path of loading on the interaction diagram, as seen in Figure 7.12.
4. Define loading step ΔGP as a small portion of the maximum load and compute the axial force at the geometric centroid:

$$\text{GP}_{new} = \text{GP}_{old} + \Delta\text{GP} \tag{7.64}$$

5. Calculate moment GM about the geometric centroid:

$$e = \frac{\text{GM}}{\text{GP}} \quad \text{and} \quad \text{GM} = e \times \text{GP} \tag{7.65}$$

6. Transfer moment to the updated inelastic centroid and calculate the new transferred moment TM, as seen in Figure 7.13:

$$\text{TM} = \text{GM} + \text{GP}(Y_G - Y_c) \tag{7.66}$$

The advantage of transferring the moment to the position of the inelastic centroid is to eliminate the coupling effect between the force and moment, since EAM = 0 about the inelastic centroid (Rasheed and Dinno 1994):

$$\begin{Bmatrix} \text{GP} \\ \text{TM} \end{Bmatrix} = \begin{bmatrix} \text{EA} & 0 \\ 0 & \text{EI} \end{bmatrix} \begin{Bmatrix} \varepsilon_0 \\ \phi \end{Bmatrix} \tag{7.67}$$

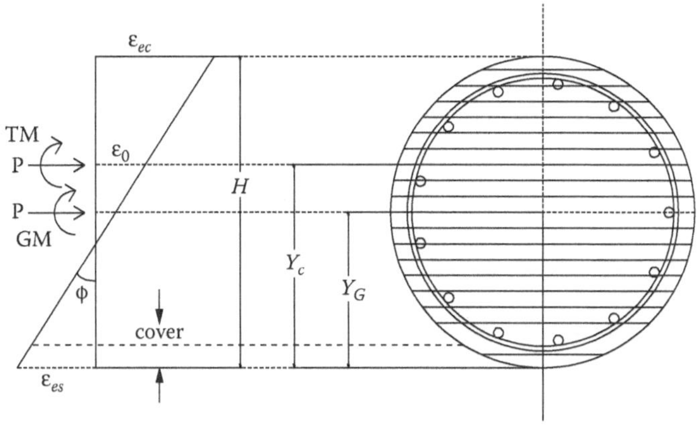

FIGURE 7.13 Transferring moment from geometric centroid to inelastic centroid.

7. Find the curvature ϕ:

$$\phi = \frac{\text{TM}}{\text{EI}} \tag{7.68}$$

Strain at the inelastic centroid ε_0, the extreme compression fiber strain ε_{ec}, and strain at the extreme level of steel in tension ε_{es} are found as follows:

$$\varepsilon_0 = \frac{\text{GP}}{\text{EA}} \tag{7.69}$$

$$\varepsilon_{ec} = \varepsilon_0 + \phi(H - Y_c) \tag{7.70}$$

$$\varepsilon_{es} = \varepsilon_0 - \phi(Y_c - \text{cover}) \tag{7.71}$$

where cover is up to the center of the bars.

8. Calculate strain ε_{ci} and corresponding stress f_{ci} in each layer of concrete section by using the selected model from step 2:

$$\varepsilon_{ci} = \varepsilon_{ec} - \phi Y_i \tag{7.72}$$

9. Calculate strain ε_{si} and corresponding stress f_{si} in each bar in the given section by using the steel model (elastic up to yield strength and then perfectly plastic):

$$\varepsilon_{si} = \varepsilon_{ec} - \phi Y_{si} \tag{7.73}$$

10. Calculating the new section properties: axial rigidity EA, flexural rigidity about the inelastic centroid EI, moment of axial rigidity about inelastic centroid EAM, internal axial force F_z, and internal bending moments about the inelastic centroid M_0,

$$EA = \sum_i E_{ci} w_i t_i + \sum_i (E_{si} - E_{ci}) A_{si} \tag{7.74}$$

$$EAM = \sum_i E_{ci} w_i t_i (H - Y_c - Y_i) + \sum_i (E_{si} - E_{ci}) A_{si} (H - Y_c - Y_{si}) \tag{7.75}$$

$$F_z = \sum_i f_{ci} w_i t_i + \sum_i (f_{si} - f_{ci}) A_{si} \tag{7.76}$$

$$EI = \sum_i E_{ci} w_i t_i (H - Y_c - Y_i)^2 + \sum_i (E_{si} - E_{ci}) A_{si} (H - Y_c - Y_{si})^2 \tag{7.77}$$

$$M_o = \sum_i f_{ci} w_i t_i (H - Y_c - Y_i) + \sum_i (f_{si} - f_{ci}) A_{si} (H - Y_c - Y_{si}) \tag{7.78}$$

where

E_{ci} = secant modulus of elasticity of the concrete layer = $\dfrac{f_{ci}}{\varepsilon_{ci}}$
E_{si} = secant modulus of elasticity of the steel bar = $\dfrac{f_{si}}{\varepsilon_{si}}$

11. Transferring back the internal moment about the geometric centroid

$$GM_0 = M_0 - GP(Y_G - Y_c) \tag{7.79}$$

12. Checking the convergence of the inelastic centroid

$$TOL = \frac{EAM}{EA \times Y_c} \tag{7.80}$$

13. Comparing the internal force to applied force, internal moments to applied moments, and ensuring that the moments are calculated about the geometric centroid

$$|GP - F_z| \leq 1 \times 10^{-5} \tag{7.81}$$

$$|GM - GM_0| \leq 1 \times 10^{-5} \tag{7.82}$$

$$|Tol| \leq 1 \times 10^{-5} \tag{7.83}$$

If Equations (7.81), (7.82), and (7.83) are not satisfied, the location of the inelastic centroid is updated by *EAM*/*EA*, and steps 5 to 11 are repeated until Equations (7.81), (7.82), and (7.83) are satisfied.

$$Y_{C_{new}} = Y_{C_{old}} + \frac{EAM}{EA} \tag{7.84}$$

Once equilibrium is reached, the algorithm checks for ultimate strain in concrete ε_{ec} and steel ε_{es} not to exceed $\overline{\varepsilon_{cu}}$ and 0.05, respectively, and then it increases the loading by ΔGP and runs the analysis for the new load level using the latest section properties. Otherwise, if ε_{ec} equals $\overline{\varepsilon_{cu}}$ or ε_{es} equals 0.05, the target force and resultant moment are reached as a point on the interaction diagram for the amount of eccentricity used. For more details about this procedure, check the work of Rasheed and Dinno (1994) and Abd El Fattah, Rasheed, and Esmaeily (2011).

Example 7.1: Analysis

One of the columns that were tested by Eid, Roy, and Paultre (2009) is investigated in this example. The following are the properties of the column:

Diameter	11.9 in.	303 mm
Clear cover	1 in.	25 mm
f'_c	4.59 ksi	31.7 MPa
$n \times t_f$	2 × 0.015 in.	2 × 0.381 mm
E_f	11,306 ksi	78,000 MPa
ε_{fu}	0.013	0.013
f_{yt}	66.1 ksi	456 MPa
Hoop spacing	3.94 in.	100 mm
Hoop diameter	#3	9.5 mm
Longitudinal bars	6 #5	6 ϕ 15.9 mm
f_y	61.3 ksi	423 MPa

Draw the interaction diagram per ACI 440.2R-08 then use KDOT Column Expert.

Solution:

Using Equation (7.7):

$$f_l = \frac{2nt_fE_f\varepsilon_{fe}}{D} = \frac{2\times2\times0.015\times11306\times0.586\times0.013}{11.9} = 0.4343\text{ksi}$$

$$\frac{f_l}{f'_c} = \frac{0.4343}{4.59} = 0.095 > 0.08, \text{ stress–strain curve is ascending}$$

Point A:

According to Equation (7.6)

$$f'_{cc} = 4.59 + 0.95 \times 3.3 \times 1 \times 0.4343 = 5.95\text{ksi}$$

$$\varepsilon_{ccu} = 0.002 \times \left(1.5 + 12 \times 1 \times 0.095 \times \left(\frac{0.586 \times 0.013}{0.002}\right)^{0.45}\right) = 0.007162$$

Following Equation (7.1) with and without ϕ factors

$$P_n = 0.85 \times 5.95 \times \left(\frac{\pi}{4} \times 11.9^2 - 6 \times 0.31\right) + 61.3 \times 6 \times 0.31 = 667.1\text{ kips}$$

$$\phi P_n = 0.75 \times 0.85 \times 667.1 = 425.3\text{ kips}$$

Point B:

$$f_l = \frac{2nt_f E_f \varepsilon_{fe}}{D} = \frac{2 \times 2 \times 0.015 \times 11{,}306 \times 0.004}{11.9} = 0.228\text{ ksi}$$

This f_l will not be used to determine whether the stress–strain curve is ascending or descending, since it is computed for eccentric points.

$$f'_{cc} = 4.59 + 0.95 \times 3.3 \times 1 \times 0.228 = 5.3\text{ ksi}$$

$$E_c = 3861.7\text{ ksi}, \qquad E_2 = 154.44\text{ ksi}, \qquad \varepsilon'_t = 0.002476$$

$$\varepsilon'_{ccu} = 0.002 \times \left[1.5 + 12 \times 1 \times 0.0497 \times \left(\frac{0.004}{0.002}\right)^{0.45}\right] = 0.004629$$

$$P_{n,B} = 483.14\text{ kips}, \qquad M_{n,B} = 53.13\text{ kip-ft}$$

Point C:

$$P_{n,C} = 286.86\text{ kips}, \qquad M_{n,C} = 78.75\text{ kip-ft}$$

Point D:

$$P_{n,D} = 0\text{ kips}, \qquad M_{n,D} = 38.54\text{ kip-ft}$$

Figure 7.14 presents the approximate interaction diagram computed according to ACI 440.2R-08 procedure along with the value of the experimental point. It is evident that the interaction diagram in Figure 7.14 is conservative with respect to the experimental point.

On the other hand, Figure 7.15 depicts the interaction diagram according to the KDOT Column Expert software. It is evident that the latter is in good agreement with the experimental point. Also, it is clear that the interaction diagram computed according to ACI 440.2R-08 is conservative with respect to that of KDOT Column Expert in the overall sense. This should be expected to happen, since the ACI440.2R-08 procedure accounts for FRP wrapping only and ignores the confinement provided by the internal steel.

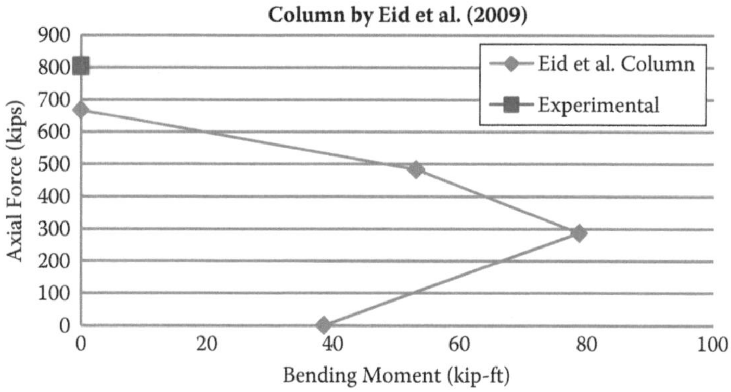

FIGURE 7.14 Confined interaction diagram according to ACI 440.2R-08 for column tested by Eid, Roy, and Paultre (2009).

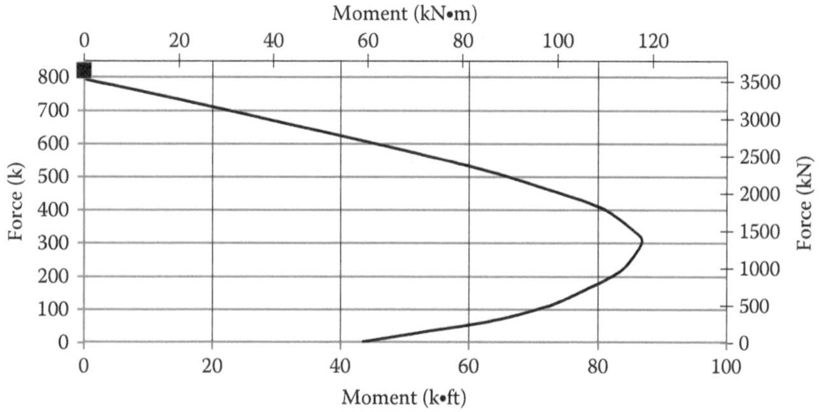

FIGURE 7.15 Confined interaction diagram according to KDOT Column Expert for column tested by Eid, Roy, and Paultre (2009).

Example 7.2: Analysis

One of the columns that were tested by Saadatmanesh, Ehsani, and Jin (1996) is investigated in this example. The following are the properties of the column:

Diameter	12 in.	305 mm
Clear cover	1 in.	25 mm
f'_c	5 ksi	34.5 MPa
$n \times t_f$	6 × 0.03 in.	6 × 0.762 mm
E_f	2,696 ksi	18,600 MPa

ε_{fu}	0.029	0.029
f_{yt}	89 ksi	614 MPa
Hoop spacing	3.5 in.	88.9 mm
Hoop diameter	#3	9.5 mm
Longitudinal bars	14#4	14 ϕ 12.7 mm
f_y	65 ksi	448 MPa

Draw the interaction diagram per ACI 440.2R-08 then use KDOT Column Expert.

Solution:

Using Equation (7.7):

$$f_l = \frac{2nt_f E_f \varepsilon_{fe}}{D} = \frac{2\times6\times0.03\times2696\times0.586\times0.029}{12} = 1.3745 \text{ ksi}$$

$$\frac{f_l}{f'_c} = \frac{1.3745}{5} = 0.275 > 0.08, \text{ stress–strain curve is ascending}$$

Point A:

According to Equation (7.6),

$$f'_{cc} = 5 + 0.95\times3.3\times1\times1.3745 = 9.31 \text{ ksi}$$

$$\varepsilon_{ccu} = 0.002\times\left[1.5+12\times1\times0.275\times\left(\frac{0.586\times0.029}{0.002}\right)^{0.45}\right] = 0.0203 > 0.01$$

$$\Rightarrow \varepsilon_{ccu\_new} = 0.01$$

Recalculate f'_{cc} based on the original E_2 and the new ε_{ccu}.

$$E_2 = \frac{f'_{cc} - f'_c}{\varepsilon_{ccu}} = \frac{9.31-5}{0.0203} = 212.315 \text{ ksi}$$

$$f'_{cc\_new} = f'_c + E_2\varepsilon_{ccu\_new} = 5 + (212.315\times0.01) = 7.123 \text{ ksi}$$

Following Equation (7.1) with and without ϕ factors,

$$P_n = 0.85\times7.123\times\left(\frac{\pi}{4}\times12^2 - 14\times0.20\right) + 65\times14\times0.20 = 849.8 \text{ kips}$$

$$\phi P_n = 0.75\times0.85\times849.8 = 541.75 \text{ kips}$$

Point B:

$$f_l = \frac{2nt_f E_f \varepsilon_{fe}}{D} = \frac{2\times6\times0.03\times2696\times0.004}{12} = 0.3235 \text{ ksi}$$

This f_l will not be used to determine whether the stress–strain curve is ascending or descending, since it is computed for eccentric points.

$$f'_{cc} = 5 + 0.95\times3.3\times1\times0.3235 = 6.014 \text{ ksi}$$

$$E_c = 4030.51 \text{ ksi}, \qquad E_2 = 198.042 \text{ ksi}, \qquad \varepsilon'_t = 0.002609$$

$$\varepsilon'_{ccu} = 0.002\times\left[1.5+12\times1\times0.0647\times\left(\frac{0.004}{0.002}\right)^{0.45}\right] = 0.005121$$

$$P_{n,B} = 591.085 \text{ kips}, \qquad M_{n,B} = 64.49 \text{ kip-ft}$$

Point C:

$$P_{n,C} = 343.01 \text{ kips}, \qquad M_{n,C} = 99.79 \text{ kip-ft}$$

Point D:

$$P_{n,D} = 0 \text{ kips}, \qquad M_{n,D} = 59.036 \text{ kip-ft}$$

By comparing Figures 7.16 and 7.17, it is evident that the interaction diagram according to ACI 440.2R-08 is conservative with respect to that of KDOT Column Expert, especially in the compression-controlled region where confinement is especially critical. This is attributed to ignoring the confinement by internal steel in ACI 440.2R-08 while it is considered by KDOT Column Expert.

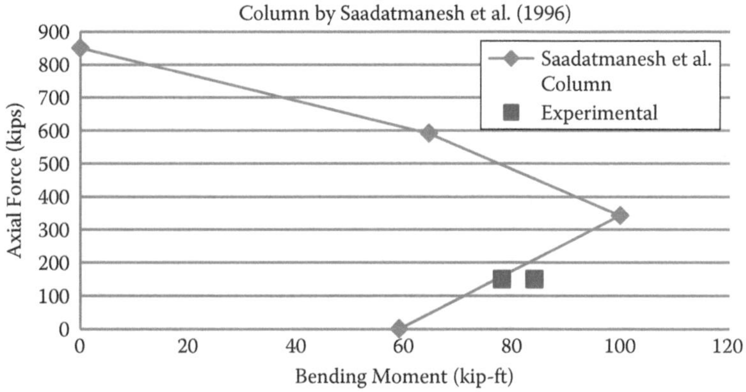

FIGURE 7.16 Confined interaction diagram according to ACI 440.2R-08 for column tested by Saadatmanesh, Ehsani, and Jin (1996).

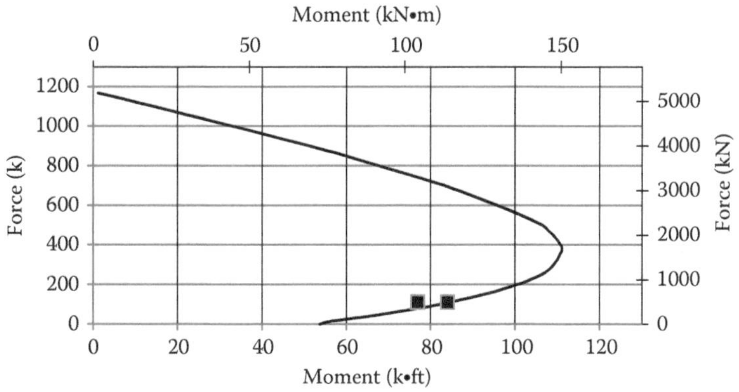

FIGURE 7.17 Confined interaction diagram according to KDOT Column Expert for column tested by Saadatmanesh, Ehsani, and Jin (1996).

7.3.3 Interaction Diagrams for Rectangular Columns

Since the derivation of the concrete contribution of force and moment equations for points B and C requires the integration of simpler integrands with a constant section width, these expressions will be evaluated analytically in closed form by integrating the stress–strain expressions perpendicular to the centroidal x-axis. The addition of the longitudinal steel bar contribution will have to be accounted for numerically anyway. However, the steel bar contribution is also easy enough to consider by hand.

7.3.3.1 Contribution of Concrete

Determining the pure axial compression point A is straightforward. To develop the equations for determining points B and C on the interaction diagram, the following expressions are derived in closed form:

$$P_{cn} = b\int_0^c f_c d_y = b\int_0^{y_t}\left[E_c\phi_s y - \frac{(E_c - E_2)^2}{4f_c'}(\phi_s y)^2\right]d_y + b\int_{y_t}^{c}\left(f_c' + E_2(\phi_s y)\right)d_y$$

$$= \left[b\phi_s E_c\frac{y^2}{2} - b\frac{(E_c - E_2)^2}{4f_c'}\phi_s^2\frac{y^3}{3}\right]_0^{y_t} + \left[bf_c'y + bE_2\phi_s\frac{y^2}{2}\right]_{y_t}^{c}$$

$$= b\phi_s\frac{E_c}{2}y_t^2 - b\frac{(E_c - E_2)^2}{12f_c'}\phi_s^2 y_t^3 + bf_c'c + b\frac{E_2}{2}c^2\phi_s - bf_c'y_t - b\frac{E_2}{2}\phi_s y_t^2$$

while the section curvature $\phi_s = \dfrac{\varepsilon_{ccu}'}{c}$, the P_{cn} expression may be written as

$$P_{cn} = \left[-b\frac{(E_c - E_2)^2}{12f_c'}\left(\frac{\varepsilon_{ccu}'}{c}\right)^2 y_t^3 + b\frac{E_c - E_2}{2}\left(\frac{\varepsilon_{ccu}'}{c}\right)y_t^2 - bf_c'y_t + bf_c'c + \frac{bE_2}{2}\varepsilon_{ccu}'c\right] \tag{7.85}$$

Similarly, the expression for M_n can be derived as follows:

$$M_{cn} = \int_0^c f_c b d_y\left(y + h - c - \frac{h}{2}\right) = \int_0^c f_c b d_y\left(y + \frac{h}{2} - c\right)$$

$$= b\int_0^{y_t}\left[E_c\phi_s y - \frac{(E_c - E_2)^2}{4f_c'}\phi_s^2 y^2\right]d_y\left(y + \frac{h}{2} - c\right) + b\int_{y_t}^{c}\left[f_c' + E_2\phi_s y\right]d_y\left(y + \frac{h}{2} - c\right)$$

$$M_{cn} = \left[bE_c\phi_s \frac{y^3}{3} + bE_c\phi_s\left(\frac{h}{2} - c\right)\frac{y^2}{2} - \frac{b(E_c - E_2)^2}{4f_c'}\phi_s^2\frac{y^4}{4} - b\frac{(E_c - E_2)^2}{4f_c'}\phi_s^2\left(\frac{h}{2} - c\right)\frac{y^3}{3}\right]_0^{y_t}$$

$$+\left[bf_c'\frac{y^2}{2} + bf_c'\left(\frac{h}{2} - c\right)y + bE_2\phi_s\frac{y^3}{3} + bE_2\phi_s\left(\frac{h}{2} - c\right)\frac{y^2}{2}\right]_{y_t}^{c}$$

$$= \frac{bE_c\phi_s}{3}y_t^3 + \frac{bE_c\phi_s}{2}\left(\frac{h}{2} - c\right)y_t^2 - \frac{b(E_c - E_2)^2}{16f_c'}\phi_s^2 y_t^4 - \frac{b(E_c - E_2)^2}{12f_c'}\phi_s^2\left(\frac{h}{2} - c\right)y_t^3$$

$$+\frac{bf_c'}{2}c^2 - \frac{bf_c'}{2}y_t^2 + bf_c'\left(\frac{h}{2} - c\right)c - bf_c'\left(\frac{h}{2} - c\right)y_t + b\frac{E_2\phi_s}{3}c^3 - b\frac{E_2\phi_s}{3}y_t^3$$

$$+ b\frac{E_2\phi_s}{2}\left(\frac{h}{2} - c\right)c^2 - b\frac{E_2\phi_s}{2}\left(\frac{h}{2} - c\right)y_t^2$$

while the section curvature $\phi_s = \frac{\varepsilon'_{ccu}}{c}$, the M_{cn} expression may be written as

$$M_{cn} = -\frac{b(E_c - E_2)^2}{16f_c'}\left(\frac{\varepsilon'_{ccu}}{c}\right)^2 y_t^4 + \frac{b(E_c - E_2)}{3}\frac{\varepsilon'_{ccu}}{c}y_t^3$$

$$-\frac{b(E_c - E_2)^2}{12f_c'}\left(\frac{\varepsilon'_{ccu}}{c}\right)^2\left(\frac{h}{2} - c\right)y_t^3 + \frac{b(E_c - E_2)}{2}\frac{\varepsilon'_{ccu}}{c}\left(\frac{h}{2} - c\right)y_t^2$$

$$-\frac{bf_c'}{2}y_t^2 - bf_c'\left(\frac{h}{2} - c\right)y_t + \frac{bE_2}{3}\varepsilon'_{ccu}c^2 + \frac{bE_2}{2}\varepsilon'_{ccu}\left(\frac{h}{2} - c\right)c$$

$$+\frac{bf_c'}{2}c^2 + bf_c'\left(\frac{h}{2} - c\right)c \tag{7.86}$$

as stated earlier, when the circular section was discussed in Section 7.3.1.1:

$$c = d \quad \text{for point B} \tag{7.87}$$

$$c = d\frac{\varepsilon'_{ccu}}{\varepsilon'_{ccu} + \varepsilon_{sy}} \quad \text{for point C} \tag{7.88}$$

$$y_t = c\frac{\varepsilon'_t}{\varepsilon'_{ccu}} \tag{7.89}$$

7.3.3.2 Contribution of Steel

The following computation steps are taken in general:

1. The depth of each bar layer measured from the top extreme fiber is

$$d_{\_bar,1} = \text{clear cover} + d_{tie} + \frac{d_{bar}}{2}$$
$$d_{\_barj} = d_{\_bar,1} + (j-1) \times \frac{d - d_{\_bar,1}}{\text{No. of layers} - 1} \tag{7.90}$$

2. The strain and stress in the steel bar layer j is

$$\varepsilon_{sj} = \frac{c - d_{\_barj}}{c} \varepsilon'_{ccu}$$

$$f_{sj} = E_s \, \varepsilon_{sj} \le f_y$$

f_{cj} is obtained from Equation (7.3) by substituting ε_{sj} for ε_c. (7.91)

3. The axial force (P_{sn}) and the bending moment (M_{sn}) contribution of steel are

$$P_{sn} = \sum_j (f_{sj} - f_{cj}) \times A_{s,barj}$$

$$M_{sn} = \sum_j (f_{sj} - f_{cj}) \times A_{s,barj} \times \left(\frac{h}{2} - d_{\_barj}\right) \tag{7.92}$$

The axial force and bending moment capacity of the section is the simple sum of the contribution of concrete and steel:

$$P_n = P_{cn} + P_{sn}$$
$$M_n = M_{cn} + M_{sn} \tag{7.93}$$

7.3.4 Interaction Diagrams for Rectangular Columns Using KDOT Column Expert

As shown in the previous section, the effective confining pressure (f_l) in the case of the two eccentric points (Point B and Point C) is lower than that of pure axial compression (Point A), where the section is fully confined. In KDOT Column Expert software, this issue of partial confinement is modeled more consistently throughout the range of eccentricities that are correlated to the ratio of the compression zone to the entire section. While the case of pure axial compression has zero eccentricity and full confinement, the pure bending case has infinite eccentricity and no confinement at all. The confined strength in between the two extremes (f'_{cc} and f'_c) is mapped gradually as a function of the compression-zone ratio (CR).

$$\bar{f}_{cc} = \frac{1}{1+\dfrac{1}{CR-0.2}} f'_{cc} + \frac{1}{0.8+CR} f'_c \tag{7.94}$$

where

$$CR = \frac{0.2\dfrac{e}{\sqrt{bh}}+0.1}{\dfrac{e}{\sqrt{bh}}} \tag{7.95}$$

The relationship in Equation (7.95) has been correlated by plotting the normalized eccentricity against the compression area to cross-sectional area ratio for rectangular cross sections having different aspect ratios at the unconfined failure level. The aspect ratios used are 1:1, 2:1, 3:1, and 4:1, as shown in Figure 7.18, selected as an example. Each curve in Figure 7.18 represents specific α angle (tan α = M_y/M_x) ranging from zero to 90°. It is seen from this figure that there is an inversely proportional relation between the normalized eccentricity and compression-zone ratio, regardless of the α angle followed. By plotting the curves from all aspect ratios into one graph and establishing the best-fit curve, Equation (7.95) is introduced.

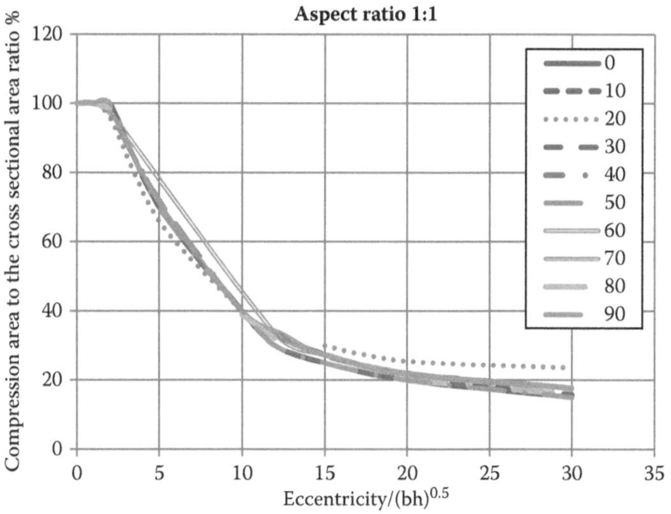

FIGURE 7.18 Compression-zone ratio vs. normalized eccentricity in rectangular columns (aspect ratio = 1:1).

It is important to note here that the eccentrically confined or partially confined Lam and Teng as well as Mander models follow the same formats described for circular columns in Sections 7.3.2.1 and 7.3.2.2.

7.3.4.1 Numerical Procedure

The column cross section is divided into a finite number of thin filaments, as seen in Figure 7.19a. The force and moment of each filament is calculated and stored. The bars are treated as discrete objects in their actual locations (Figure 7.19b). The advantage of that is to precisely calculate the internal forces induced by steel bars and concrete filaments in the column cross section. The cross section analyzed is loaded incrementally by maintaining a certain eccentricity between the axial force P and the resultant moment M_R. Since increasing the load and resultant moment causes the neutral axis and centroid to vary nonlinearly, the generalized moment-of-area theorem is devised. The method is developed using an incremental iterative analysis algorithm, a secant stiffness approach, and proportional or radial loading. It is explained in the following steps:

1. Calculating the initial section properties:
   Elastic axial rigidity EA:

$$EA = \sum_i E_c w_i t_i + \sum_i (E_s - E_c) A_{si} \tag{7.96}$$

where

E_c = initial modulus of elasticity of the concrete
E_s = initial modulus of elasticity of the steel bar

The depth of the elastic centroid position from the bottom fiber of the section Y_c and from the left side of the section X_c

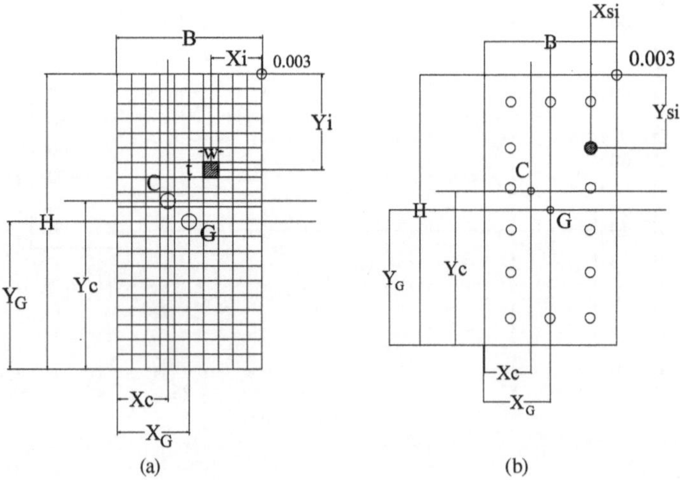

FIGURE 7.19 Geometric properties of concrete filaments and steel rebars.

$$Y_c = \frac{\sum_i E_c w_i t_i (H - Y_i) + \sum_i (E_s - E_c) A_{si} (H - Y_{si})}{EA} \tag{7.97}$$

$$X_c = \frac{\sum_i E_c w_i t_i (B - X_i) + \sum_i (E_s - E_c) A_{si} (B - X_{si})}{EA} \tag{7.98}$$

where Y_i and Y_{si} are measured from the top extreme fiber, and X_i and X_{si} are measured from the rightmost extreme fiber, as seen in Figure 7.19.
Elastic flexural rigidities about the elastic centroid EI_X, EI_y and EI_{xy}

$$EI_x = \sum_i E_c w_i t_i (H - Y_i - Y_c)^2 + \sum_i (E_s - E_c) A_{si} (H - Y_{si} - Y_c)^2 \tag{7.99}$$

$$EI_y = \sum_i E_c w_i t_i (B - X_i - X_c)^2 + \sum_i (E_s - E_c) A_{si} (B - X_{si} - X_c)^2 \tag{7.100}$$

$$EI_{xy} = \sum_i E_c w_i t_i (H - Y_i - Y_c)(B - X_i - X_c) + \sum_i (E_s - E_c) A_{si} (H - Y_{si} - Y_c)(B - X_{si} - X_c) \tag{7.101}$$

Typically the initial elastic $Y_c = H/2$, $X_c = B/2$, and $EI_{xy} = 0$

The depth of the geometric section centroid position from the bottom and left fibers of the section Y_G, X_G is

$$Y_G = \frac{H}{2} \tag{7.102}$$

$$X_G = \frac{B}{2} \tag{7.103}$$

2. Defining the eccentricity e, which specifies the radial path of loading on the interaction diagram (Figure 7.20), and also defining the angle α in between the resultant moment GM_R and GM_X
3. Defining the loading step ΔGP as a small portion of the maximum load, and computing the axial force at the geometric centroid,

$$GP_{new} = GP_{old} + \Delta GP \tag{7.104}$$

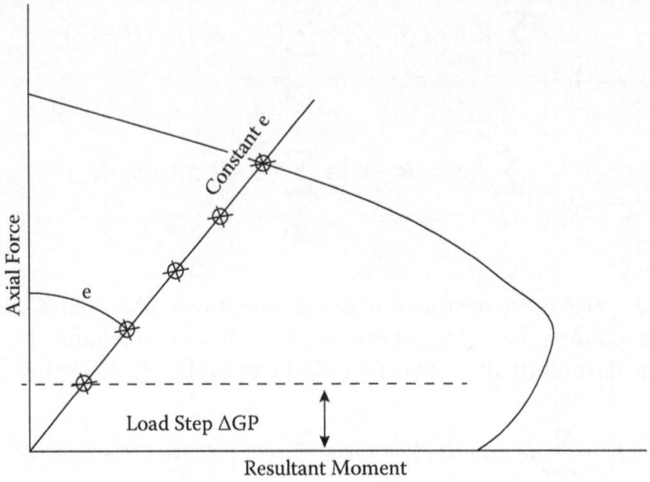

FIGURE 7.20 Radial-loading concept.

4. Calculating the moment GM_R about the geometric centroid.

$$e = \frac{GM_R}{GP} \quad \text{and} \quad GM_R = e \times GP \tag{7.105}$$

$$GM_x = GM_R \cos\alpha \tag{7.106}$$

$$GM_y = GM_x \tan\alpha \tag{7.107}$$

5. Transferring the moments to the inelastic centroid and calculating the new transferred moments TM_X and TM_Y

$$TM_x = GM_x + GP(Y_G - Y_c) \tag{7.108}$$

$$TM_y = GM_y + GP(X_G - X_c) \tag{7.109}$$

The advantage of transferring the moment to the position of the inelastic centroid is to eliminate the coupling effect between the force and the two moments, since $EAM_x = EAM_y = 0$ about the inelastic centroid (Figure 7.21).

$$\begin{bmatrix} GP \\ TM_x \\ TM_y \end{bmatrix} = \begin{bmatrix} EA & 0 & 0 \\ 0 & EI_x & EI_{xy} \\ 0 & EI_{xy} & EI_y \end{bmatrix} \begin{bmatrix} \varepsilon_0 \\ \phi_x \\ \phi_y \end{bmatrix} \tag{7.110}$$

6. Finding curvatures ϕ_x and ϕ_y by inverting Equation (7.110):

$$\phi_x = \frac{TM_x}{\beta^2} \times EI_y - \frac{TM_y}{\beta^2} \times EI_{xy} \tag{7.111}$$

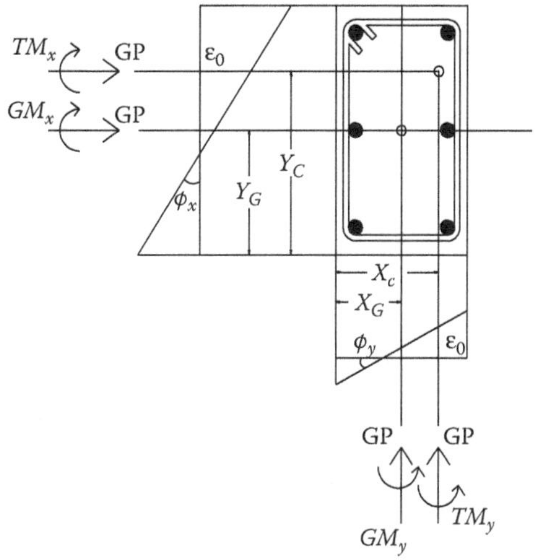

FIGURE 7.21 Moment transferring from geometric centroid to inelastic centroid.

$$\phi_y = \frac{TM_y}{\beta^2} \times EI_x - \frac{TM_x}{\beta^2} \times EI_{xy} \tag{7.112}$$

$$\beta^2 = EI_x EI_y - EI_{xy}^2 \tag{7.113}$$

7. Strain at the inelastic centroid ε_0, the extreme compression fiber strain ε_{ec}, and strain at the extreme level of steel in tension ε_{es} are found as follows:

$$\varepsilon_0 = \frac{GP}{EA} \tag{7.114}$$

$$\varepsilon_{ec} = \varepsilon_0 + \phi_x(H - Y_c) + \phi_y(B - X_c) \tag{7.115}$$

$$\varepsilon_{es} = \varepsilon_0 - \phi_x(Y_c - \text{Cover}) - \phi_y(X_c - \text{Cover}) \tag{7.116}$$

where the cover is up to the center of the bars

8. Calculating strain ε_{ci} and corresponding stress f_{ci} in each filament of concrete section by using the Lam and Teng or the Mander model.

$$\varepsilon_{ci} = \frac{GP}{EA} + \frac{TM_x(H - Y_c - Y_i)}{\beta^2} EI_y + \frac{TM_y(B - X_c - X_i)}{\beta^2} EI_x - \frac{TM_x(B - X_c - X_i)}{\beta^2} EI_{xy} - \frac{TM_y(H - Y_c - Y_i)}{\beta^2} EI_{xy} \tag{7.117}$$

9. Calculating strain ε_{si} and corresponding stress f_{si} in each bar in the given section by using the steel material model of elastic–perfectly plastic.

$$\varepsilon_{si} = \frac{GP}{EA} + \frac{TM_x(H - Y_c - Y_{si})}{\beta^2} EI_y + \frac{TM_y(B - X_c - X_{si})}{\beta^2} EI_x$$

$$- \frac{TM_x(B - X_c - X_{si})}{\beta^2} EI_{xy} - \frac{TM_y(H - Y_c - Y_{si})}{\beta^2} EI_{xy} \quad (7.118)$$

10. Calculating the new section properties: axial rigidity EA; flexural rigidities about the inelastic centroid EI_x, EI_y, EI_{xy}; moment of axial rigidity about inelastic centroid EAM_x, EAM_y; internal axial force F_z; and internal bending moments about the inelastic centroid M_{0x}, M_{0y}:

$$EA = \sum_i E_{ci} w_i t_i + \sum_i (E_{si} - E_{ci}) A_{si} \quad (7.119)$$

$$EAM_x = \sum_i E_{ci} w_i t_i (H - Y_c - Y_i) + \sum_i (E_{si} - E_{ci}) A_{si} (H - Y_c - Y_{si}) \quad (7.120)$$

$$EAM_y = \sum_i E_{ci} w_i t_i (B - X_c - X_i) + \sum_i (E_{si} - E_{ci}) A_{si} (B - X_c - X_{si}) \quad (7.121)$$

$$F_z = \sum_i f_{ci} w_i t_i + \sum_i (f_{si} - f_{ci}) A_{si} \quad (7.122)$$

$$EI_x = \sum_i E_{ci} w_i t_i (H - Y_c - Y_i)^2 + \sum_i (E_{si} - E_{ci}) A_{si} (H - Y_c - Y_{si})^2 \quad (7.123)$$

$$EI_y = \sum_i E_{ci} w_i t_i (B - X_c - X_i)^2 + \sum_i (E_{si} - E_{ci}) A_{si} (B - X_c - X_{si})^2 \quad (7.124)$$

$$EI_{xy} = \sum_i E_{ci} w_i t_i (H - Y_c - Y_i)(B - X_c - X_i)$$

$$+ \sum_i (E_{si} - E_{ci}) A_{si} (H - Y_c - Y_{si})(B - X_c - X_{si}) \quad (7.125)$$

$$M_{0x} = \sum_i f_{ci} w_i t_i (H - Y_c - Y_i) + \sum_i (f_{si} - f_{ci}) A_{si} (H - Y_c - Y_{si}) \quad (7.126)$$

$$M_{0y} = \sum_i f_{ci} w_i t_i (B - X_c - X_i) + \sum_i (f_{si} - f_{ci}) A_{si} (B - X_c - X_{si}) \quad (7.127)$$

where

E_{ci} = secant modulus of elasticity of the concrete filament = $\frac{f_{ci}}{\varepsilon_{ci}}$

E_{si} = secant modulus of elasticity of the steel bar = $\frac{f_{si}}{\varepsilon_{si}}$

11. Transferring back the internal moment about the geometric centroid

$$GM_{0x} = M_{0x} - GP(Y_G - Y_c) \quad (7.128)$$

$$GM_{0y} = M_{0y} - GP(X_G - X_c) \quad (7.129)$$

12. Checking the convergence of the inelastic centroid

$$TOL_x = EAM_x / (EA \times Y_c) \quad (7.130)$$

$$TOL_y = EAM_y / (EA \times X_c) \quad (7.131)$$

13. Comparing the internal force to applied force and the internal moments to applied moments, and making sure the inelastic centroid converges:

$$|GP - F_z| \leq 1 \times 10^{-5} \quad (7.132)$$

$$|GM_x - GM_{0x}| \leq 1 \times 10^{-5} \text{ and } |GM_y - GM_{0y}| \leq 1 \times 10^{-5} \quad (7.133)$$

$$|TOL_x| \leq 1 \times 10^{-5} \text{ and } |TOL_y| \leq 1 \times 10^{-5} \quad (7.134)$$

If Equations (7.132), (7.133), and (7.134) are not satisfied, the location of the inelastic centroid is updated by EAM_x/EA and EAM_y/EA, and steps 5 to 12 are repeated until Equations (7.132), (7.133), and (7.134) are satisfied.

$$Y_{C_{new}} = Y_{C_{old}} + \frac{EAM_x}{EA} \quad (7.135)$$

$$X_{C_{new}} = X_{C_{old}} + \frac{EAM_y}{EA} \quad (7.136)$$

Once equilibrium is reached, the algorithm checks for ultimate strain in concrete ε_{ec} and steel ε_{es} not to exceed $\overline{\varepsilon_{ccu}}$ and 0.05, respectively, and then it increases the loading by ΔGP and runs the analysis for the new load level using the latest section properties. Otherwise, if ε_{ec} equals $\overline{\varepsilon_{ccu}}$ or ε_{es} equals 0.05, the target force and resultant moment are reached as a point on the interaction diagram for the amount of eccentricity, and angle α used.

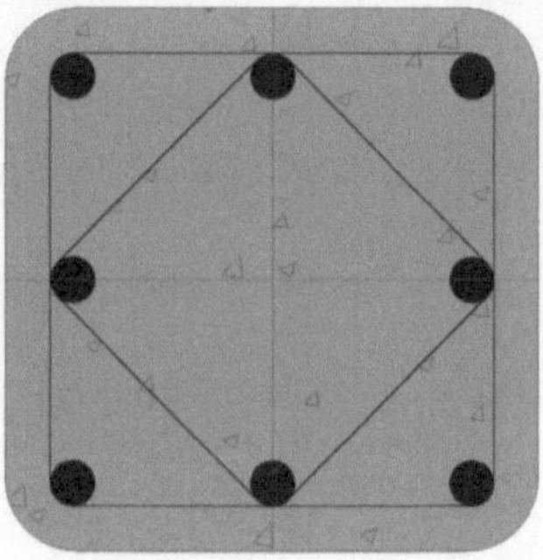

FIGURE 7.22 Section of column tested by Memon and Sheikh (2005).

Example 7.3: Analysis

One of the columns that were tested by Memon and Sheikh (2005) is investigated in this example. The following are the properties of the column:

Height	12 in.	305 mm
Width	12 in.	305 mm
Clear cover	1 in.	25 mm
r_c	0.64 in.	16.26 mm
f'_c	6.19 ksi	42.7 MPa
$n \times t_f$	4 × 0.05 in.	4 × 1.27 mm
E_f	2,864.33 ksi	19,761 MPa
ε_{fu}	0.0228	0.0228
f_{yt}	66.63 ksi	459.7 MPa
Hoop spacing	11.8 in.	299.7 mm
Hoop diameter	#3	9.5 mm
Longitudinal bars	8 #6 (ϕ = 0.77 in.)	8 ϕ 19.6 mm
f_y	67.43 ksi	465.2 MPa

Draw the interaction diagram per ACI 440.2R-08 then use KDOT Column Expert.

Solution:

Using Equation (7.7):

$$f_l = \frac{2nt_fE_f\varepsilon_{fe}}{D} = \frac{2\times4\times0.05\times2864.33\times0.586\times0.0228}{\sqrt{12^2+12^2}} = 0.902\text{ksi}$$

$$\frac{f_l}{f'_c} = \frac{0.902}{6.19} = 0.146 > 0.08$$, stress–strain curve is ascending

Point A:

$$\rho_s = \frac{8 \times \pi/4(0.77)^2}{12 \times 12} = 0.0259$$

$$\frac{A_e}{A_c} = \frac{1 - \dfrac{\left[\left(\dfrac{b}{h}\right)(h-2r_c)^2 + \left(\dfrac{h}{b}\right)(b-2r_c)^2\right]}{3A_g} - \rho_s}{1-\rho_s} = \frac{1 - \dfrac{[2\times(12-2\times0.64)^2]}{3\times144} - 0.0259}{1-0.0259}$$

$$= 0.454$$

$$\kappa_a = \frac{A_e}{A_c}\left(\frac{b}{h}\right)^2 = 0.454$$

$$\kappa_b = \frac{A_e}{A_c}\left(\frac{h}{b}\right)^{0.5} = 0.454$$

According to Equation (7.6):

$$f'_{cc} = 6.19 + 0.95 \times 3.3 \times 0.454 \times 0.902 = 7.47 \text{ ksi}$$

$$\varepsilon_{ccu} = 0.002 \times \left[1.5 + 12 \times 0.454 \times 0.146 \times \left(\frac{0.586 \times 0.0228}{0.002}\right)^{0.45}\right] = 0.00674$$

Following Equation (7.2) with and without ϕ factors:

$$P_n = 0.85 \times 7.47 \times (12 \times 12 - 8 \times \frac{\pi}{4} \times 0.77^2) + 67.43 \times 8 \times \frac{\pi}{4} \times 0.77^2 = 1141.87 \text{ kips}$$

$$\phi P_n = 0.65 \times 0.8 \times 1141.87 = 593.77 \text{ kips}$$

Point B:

$$f_l = \frac{2nt_f E_f \varepsilon_{fe}}{D} = \frac{2 \times 4 \times 0.05 \times 2864.33 \times 0.004}{\sqrt{12^2 + 12^2}} = 0.270 \text{ ksi}$$

This f_l will not be used to determine whether the stress–strain curve is ascending or descending since it is computed for eccentric points.

$$f'_{cc} = 6.19 + 0.95 \times 3.3 \times 0.454 \times 0.270 = 6.574 \text{ ksi}$$

$$E_c = 4484.56 \text{ ksi}, \quad E_2 = 105.323 \text{ ksi}, \quad \varepsilon'_t = 0.002827$$

$$\varepsilon'_{ccu} = 0.002 \times \left[1.5 + 12 \times 0.454 \times \frac{0.27}{6.19} \times \left(\frac{0.004}{0.002}\right)^{0.45}\right] = 0.003649$$

$$P_{n,B} = 709.18 \text{ kips}, \quad M_{n,B} = 127.90 \text{ kip-ft}$$

Point C:

$$P_{n,C} = 353.5 \text{ kips}, \quad M_{n,C} = 169.8 \text{ kip-ft}$$

Point D:

$$P_{n,D} = 0 \text{ kips}, \quad M_{n,D} = 97.37 \text{ kip-ft}$$

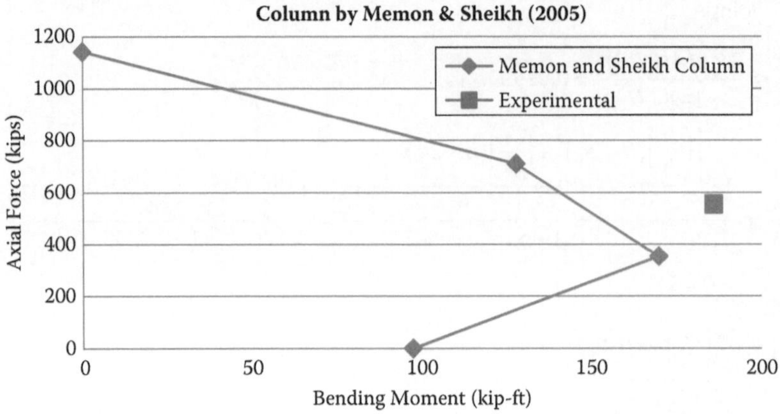

FIGURE 7.23 Confined interaction diagram according to ACI 440.2R-08 for column MS3 tested by Memon and Sheikh (2005).

By comparing Figures 7.23 and 7.24, it is evident that the ACI 440.2R-08 procedure is conservative compared to the curve generated by the KDOT Column Expert software. This is again attributed to the fact that the ACI 440.2R-08 procedure ignores the confinement resulting from internal reinforcement, which is significant due to the extra diamond-shaped tie that is accounted for by the KDOT Column Expert software.

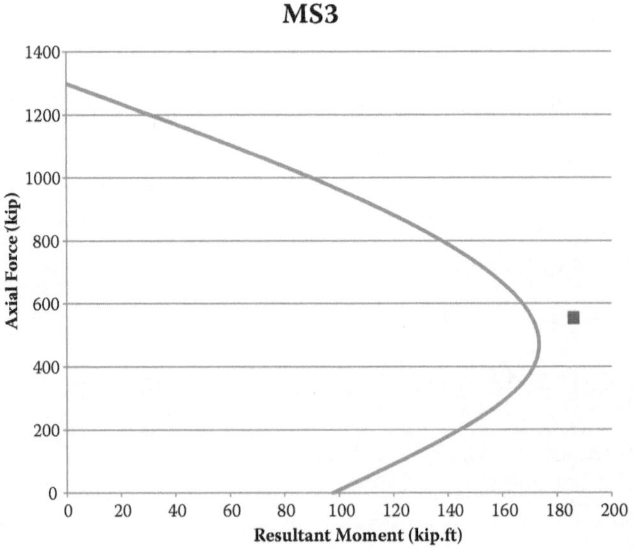

FIGURE 7.24 Confined interaction diagram according to KDOT Column Expert for column MS3 tested by Memon and Sheikh (2005).

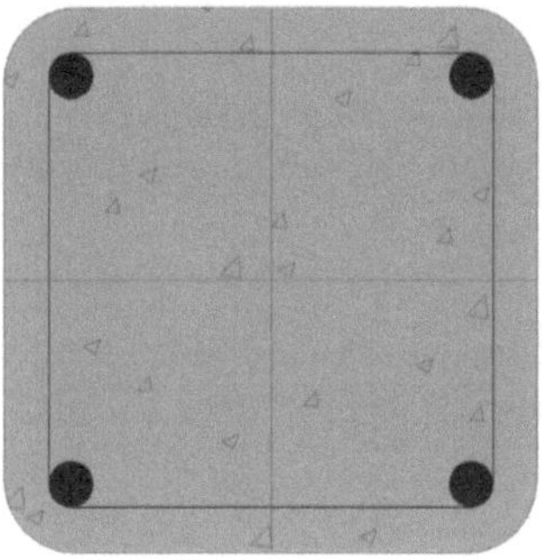

FIGURE 7.25 Section of column SC2 tested by Darby et al. (2011).

Example 7.4: Analysis

One of the columns that were tested by Darby et al. (2011) is investigated in this example (Figure 7.25). The following are the properties of the column:

Height	11.81 in.	300 mm
Width	11.81 in.	300 mm
Clear cover	1 in.	25 mm
r_c	1.57 in.	39.88 mm
f'_c	3.625 ksi	25.0 MPa
$n \times t_f$	4 × 0.0063 in.	4 × 0.16 mm
E_f	31,030 ksi	214.08 GPa
ε_{fu}	0.0145	0.0145
f_{yt}	60 ksi	413.93 MPa
Tie spacing	11 in.	279.4 mm
Tie diameter	#3	9.5 mm
Longitudinal bars	4 #8 (ϕ = 0.98 in.)	4 ϕ 24.9 mm
f_y	79.75 ksi	550.2 MPa

Draw the interaction diagram per ACI 440.2R-08 then use KDOT Column Expert.

Solution:

Using Equation (7.7):

$$f_l = \frac{2nt_f E_f \varepsilon_{fe}}{D} = \frac{2\times 4\times 0.0063\times 31030\times 0.586\times 0.0145}{\sqrt{11.81^2+11.81^2}} = 0.796\text{ksi}$$

$\frac{f_l}{f'_c} = \frac{0.796}{3.625} = 0.219 > 0.08$, stress–strain curve is ascending

Point A:

$$\rho_s = \frac{4\times\pi/4(0.98)^2}{11.81\times 11.81} = 0.0216$$

$$\frac{A_e}{A_c} = \frac{1 - \dfrac{\left[\left(\dfrac{b}{h}\right)(h-2r_c)^2 + \left(\dfrac{h}{b}\right)(b-2r_c)^2\right]}{3A_g} - \rho_s}{1-\rho_s}$$

$$= \frac{1 - \dfrac{[2\times(11.81-2\times 1.57)^2]}{3\times 11.81\times 11.81} - 0.0216}{1-0.0216} = 0.633$$

$$\kappa_a = \frac{A_e}{A_c}\left(\frac{b}{h}\right)^2 = 0.633$$

$$\kappa_b = \frac{A_e}{A_c}\left(\frac{h}{b}\right)^{0.5} = 0.633$$

According to Equation (7.6),

$$f'_{cc} = 3.625 + 0.95\times 3.3\times 0.633\times 0.796 = 5.2\text{ ksi}$$

$$\varepsilon_{ccu} = 0.002\times\left(1.5 + 12\times 0.633\times 0.219\times\left[\frac{0.586\times 0.0145}{0.002}\right]^{0.45}\right) = 0.00938$$

Following Equation (7.2) with and without ϕ factors,

$$P_n = 0.85\times 5.2\times(11.81\times 11.81 - 4\times\frac{\pi}{4}\times 0.98^2) + 79.75\times 4\times\frac{\pi}{4}\times 0.98^2 = 843.8\text{ kips}$$

$$\phi P_n = 0.65\times 0.8\times 1141.87 = 438.76\text{ kips}$$

Point B:

$$f_l = \frac{2nt_f E_f \varepsilon_{fe}}{D} = \frac{2\times 4\times 0.0063\times 31{,}030\times 0.004}{\sqrt{11.81^2 + 11.81^2}} = 0.375\text{ ksi}$$

This f_l will not be used to determine whether the stress–strain curve is ascending or descending, since it is computed for eccentric points.

$$f'_{cc} = 3.625 + 0.95\times 3.3\times 0.633\times 0.375 = 4.368\text{ ksi}$$

$$E_c = 3431.85\text{ ksi},\qquad E_2 = 144.49\text{ ksi},\qquad \varepsilon'_t = 0.002205$$

$$\varepsilon'_{ccu} = 0.002\times\left[1.5 + 12\times 0.633\times\frac{0.375}{3.625}\times\left(\frac{0.004}{0.002}\right)^{0.45}\right] = 0.005147$$

$$P_{n,B} = 522.45\text{ kips},\qquad M_{n,B} = 95.93\text{ kip-ft}$$

Point C:

$$P_{n,C} = 259.97\text{ kips},\qquad M_{n,C} = 148.96\text{ kip-ft}$$

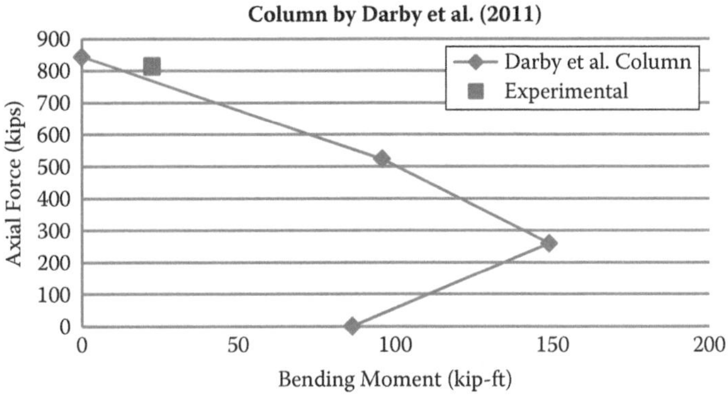

FIGURE 7.26 Confined interaction diagram according to ACI 440.2R-08 for column SC2 tested by Darby et al. (2011).

Point D:

$P_{n,D} = 0$ kips, $M_{n,D} = 86.375$ kip-ft

When comparing Figures 7.26 and 7.27, it is evident again that the ACI 440.2R-08 procedure is more conservative compared to that of KDoT Column Expert procedure.

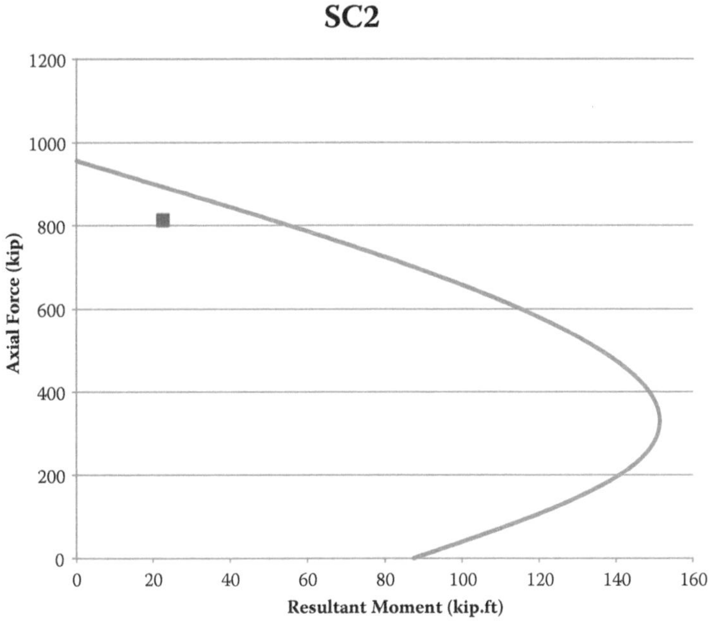

FIGURE 7.27 Confined interaction diagram according to KDOT Column Expert for column SC2 tested by Darby et al. (2011).

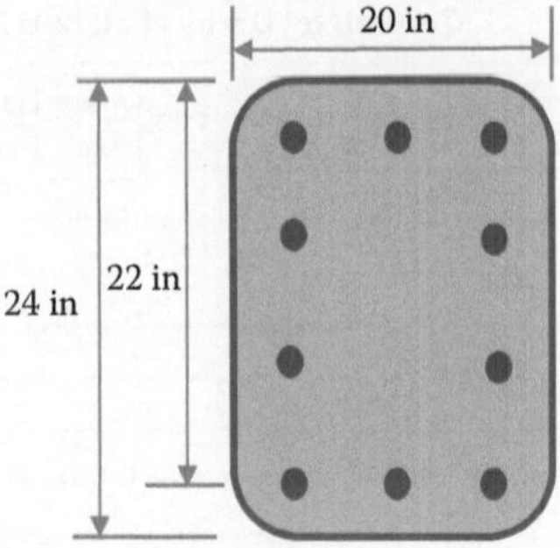

FIGURE 7.28 Column section of example 7.5.

Example 7.5: Design

This example is intended to increase the axial load capacity of a rectangular column by 20% by means of increasing confinement. The following data describe the column, Figure 7.28.

Height	24 in.	610 mm
Width	20 in.	508 mm
Clear cover	1 in.	25 mm
r_c	1.2 in.	30.5 mm
f'_c	6.0 ksi	41.39 MPa
t_f	0.015 in.	0.381 mm
E_f	33,000 ksi	227.67 GPa
ε_{fu}	0.0167	0.0167
f_{yt}	60 ksi	413.93 MPa
Tie spacing	18 in.	457.2 mm
Tie diameter	#3	9.5 mm
Longitudinal bars	10 #10	10 ϕ 32 mm
f_y	60 ksi	413.93 MPa

Solution:

$$\varepsilon_{fu} = C_E \varepsilon^*_{fu} = 0.95 \times 0.0167 = 0.0159$$

$$\phi P_{existing} = 0.8 \times 0.65[0.85 \times 6 \times (24 \times 20 - 10 \times 1.27) + 60 \times 10 \times 1.27] = 1635.52 \text{ kips}$$

$$\phi P_{nreq'd} = 1635.52\ k \times 1.2 = 1962.62\text{k}$$

$$= 0.8\phi\left[0.85 f'_{cc}(A_g - A_{st}) + A_{st} f_y\right] = 0.8 \times 0.65 \times \left[0.85 \times f'_{cc}(24 \times 20 - 12.7)\right.$$
$$\left. + 12.7 \times 60\right]$$

$$f'_{cc} = 7.584 \text{ ksi}$$

$$f'_{cc} = f'_c + \Psi_f \times 3.3 \times k_a f_l$$

$$k_a = \frac{A_e}{A_c}\left(\frac{b}{h}\right)^2 = 0.458(0.833)^2 = 0.318$$

$$\rho_s = \frac{10 \times 1.27}{480} = 0.0265$$

$$\frac{A_e}{A_c} = \frac{1 - \dfrac{\left[(0.833)(24 - 2\times 1.2)^2 + (1.2)(20 - 2\times 1.2)^2\right]}{3\times 480} - 0.0265}{1 - 0.0265} = 0.458$$

$$7.584 = 6.0 + 0.95 \times 3.3 \times 0.318 f_l$$

$$f_l = 1.589 \text{ ksi}$$

$$f_l = \frac{2nt_f E_f \varepsilon_{fe}}{\sqrt{b^2 + h^2}}$$

$$1.589 = \frac{2n \times 0.015 \times 33{,}000 \times 0.586 \times 0.0159}{\sqrt{24^2 + 20^2}}$$

$n = 5.38$ plies, use 6 plies

$$\frac{f_l}{f'_c} = \frac{1.589}{6.0} = 0.265 > 0.08$$

$$k_b = \frac{A_e}{A_c}\left(\frac{h}{b}\right)^{0.5} = 0.458\sqrt{1.2} = 0.502$$

$$\varepsilon_{ccu} = \varepsilon'_c\left(1.5 + 12k_b \frac{f_l}{f'_c}\left(\frac{\varepsilon_{fe}}{\varepsilon'_c}\right)^{0.45}\right) \le 0.01$$

$$\varepsilon'_c = 1.71\frac{f'_c}{E_c} = 3\times 10^{-5}\sqrt{f'_c} = 0.00232$$

$$\varepsilon_{ccu} = 0.00232\left[1.5 + 12 \times 0.502 \times 0.265\left(\frac{0.586 \times 0.0159}{0.00232}\right)^{0.45}\right] = 0.0104 > 0.01$$

f'_{cc} needs to be adjusted to correspond to 0.01.
Recalculate f'_{cc} based on the original E_2 and the new ε_{ccu}.

$$E_2 = \frac{f'_{cc} - f'_c}{\varepsilon_{ccu}} = \frac{7.584 - 6}{0.0104} = 152.26 \text{ ksi}$$

$$f'_{cc\_new} = f'_c + E_2\varepsilon_{ccu\_new} = 6 + 152.26 \times 0.01 = 7.523 \text{ ksi}$$

Example 7.6: Analysis

Determine the compression-controlled interaction diagram for the column in Example 7.5 without FRP reinforcement. Use the simplified calculation approach. Scale the ultimate unconfined interaction diagram to the design diagram using the appropriate ϕ factors.

Solution:

There are three points to determine for the compression-controlled interaction diagram:

1. Point A: Pure compression

$$P_{nA} = 0.85f_c'(A_g - A_{st}) + A_{st}f_y = 0.85 \times 6.0 \times (480 - 10 \times 1.27)$$
$$+ 10 \times 1.27 \times 60 = 3145.23\ k$$

$$\phi P_{nA} = 0.65 \times 0.8 \times 3145.23 = 1635.52\ k$$

2. Point B: Compression + bending when the extreme steel layer has a zero strain (Figure 7.29)

$$P_n = 0.85f_c'ba + A_{s1}f_y + A_{s2}f_{s2} + A_{s3}f_{s3}$$

$$P_{nB} = 0.85 \times 6.0 \times 20 \times 0.75 \times 22 + 3 \times 1.27 \times 60 + 2 \times 1.27 \times 29{,}000 \times 0.00182$$
$$+ 2 \times 1.27 \times 29{,}000 \times 0.000909 = 2112.62$$

$$c = d = 22''$$

$$a = \beta_1 c = \left(1.05 - 0.05\frac{6000}{1000}\right) \times 22 = 0.75 \times 22'' = 16.5''$$

$$\phi P_{nB} = 0.65 \times 2112.62 = 1373.2\ \text{k}$$

$$P_n \times e' = 0.85f_c'ba\left(d - \frac{a}{2}\right) + A_{s1}f_y\,(d - d') + A_{s2}f_{s2}\left(\frac{2}{3}(d - d')\right)$$
$$+ A_{s3}f_{s3}\left(\frac{d - d'}{3}\right)$$

$$P_n \times e_B' = 0.85 \times 6.0 \times 20 \times 16.5\left(22 - \frac{16.5}{2}\right) + 3.81 \times 60 \times 20$$
$$+ 2.54 \times 29{,}000 \times 0.00182 \times \frac{2}{3} \times 20$$
$$+ 2.54 \times 29{,}000 \times 0.000909 \times \frac{20}{3} = 29{,}947.11\ \text{k-in.}$$

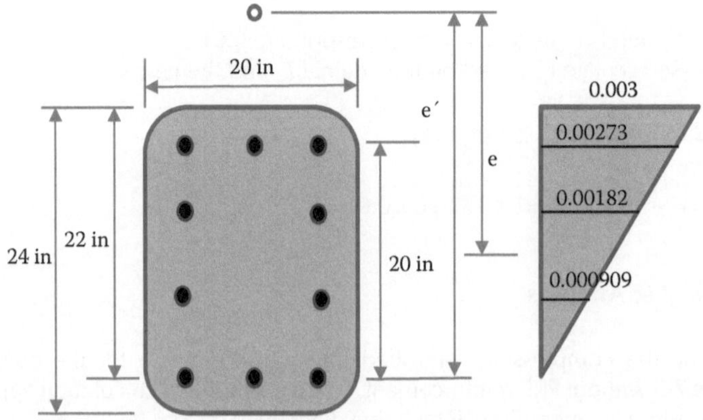

FIGURE 7.29 Determination of Point B on the unconfined interaction diagram.

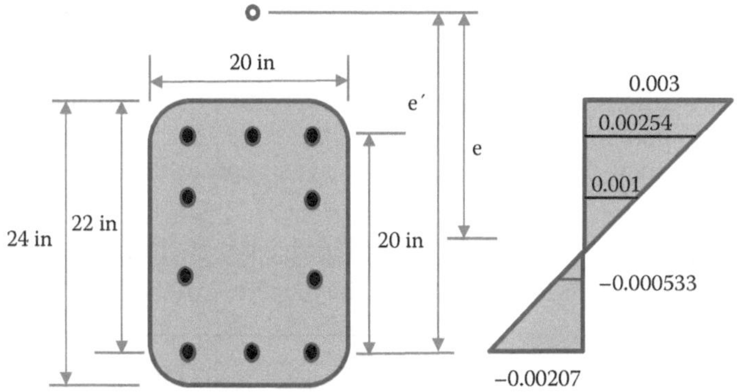

FIGURE 7.30 Determination of Point C on the unconfined interaction diagram.

$$2112.62\left(e+\frac{h}{2}-d'\right)=29{,}947.11$$

$$e=4.175''$$

$$\Rightarrow M_{nB}=2112.62\times 4.175=8820.91\text{ k-in.}=735.1\text{ k-ft}$$

$$\phi M_{nB}=0.65\times 735.1=477.8\text{ k-ft}$$

3. Point C: Compression + bending when the balanced behavior is observed (Figure 7.30)

$$\frac{0.003}{c}=\frac{0.00507}{22}\Rightarrow c=13.02''$$

$$a=0.75\times c=9.765''$$

$$\begin{aligned}P_{nC}&=0.85\times 6.0\times 20\times 9.765+3.81\times 60+2.54\times 29{,}000\times 0.001\\&\quad-2.54\times 29{,}000\times 0.000533\\&=1259.03\text{ k}\end{aligned}$$

$$\phi P_{nC}=818.4\text{ k}$$

$$\begin{aligned}P_n\times e_c'&=0.85\times 6.0\times 20\times 9.765\times\left(22-\frac{9.765}{2}\right)+3.81\times 60\times 20\\&\quad+2.54\times 29{,}000\times 0.001\left(\frac{2}{3}\times 20\right)\\&\quad-2.54\times 29{,}000\times 0.000533\times\left(\frac{20}{3}\right)\\&=22{,}341.94\end{aligned}$$

$$1259.03\times(e+12-2)=22{,}341.94$$

$$e=7.745''$$

$$M_{nC}=1259.03\times 7.745=9751.64\text{ k-in.}=812.64\text{ k-ft}$$

$$\phi M_{nC}=528.21\text{ k-ft}$$

Chapter Problems

Problem 7.1

For the rectangular column designed in Example 7.5 with 6 FRP plies for confinement, plot the approximate confined interaction diagram and compare it to the plot of Example 7.6 to show the improvement in behavior between the two cases.

Problem 7.2

One of the circular columns that were tested by Eid, Roy, and Paultre (2009) is investigated in this problem. The following are the properties of the column:

Diameter	11.9 in.	303 mm
Clear cover	1 in.	25 mm
f'_c	7.35 ksi	50.7 MPa
$n \times t_f$	2 × 0.015 in.	2 × 0.381 mm
E_f	11,306 ksi	78,000 MPa
ε_{fu}	0.013	0.013
f_{yt}	66.1 ksi	456 MPa
Hoop spacing	2.56 in.	65 mm
Hoop diameter	#3	9.5 mm
Longitudinal bars	6 #5	6 ϕ 15.9 mm
f_y	61.3 ksi	423 MPa

Plot the approximate confined interaction diagram. The experimental pure axial compression point is at $P_n = 1250$ kips. Note that this column is similar to that of Example 7.1 except for different f'_c and spiral spacing.

Problem 7.3

One of the circular columns that were tested by Sheikh and Yau (2002) is investigated in this problem. The following are the properties of the column:

Diameter	14 in.	355 mm
Clear cover	1 in.	25 mm
f'_c	6.49 ksi	44.8 MPa
$n \times t_f$	1 × 0.039 in.	1 × 0.99 mm
E_f	10,871 ksi	75,000 MPa
ε_{fu}	0.013	0.013
f_{yt}	72.5 ksi	500 MPa
Hoop spacing	11.8 in.	300 mm
Hoop diameter	#3	9.5 mm
Longitudinal bars	6 #8	6 ϕ 25.4 mm
f_y	72.5 ksi	500 MPa

Plot the approximate confined interaction diagram. The experimental near-balanced point is at $P_n = 325$ kips and $M_n = 200$ k-ft.

Problem 7.4

One of the rectangular columns (B03-b) that was tested by Bousias et al. (2004) is investigated in this problem. The following are the properties of the column, see Figure 7.P.1:

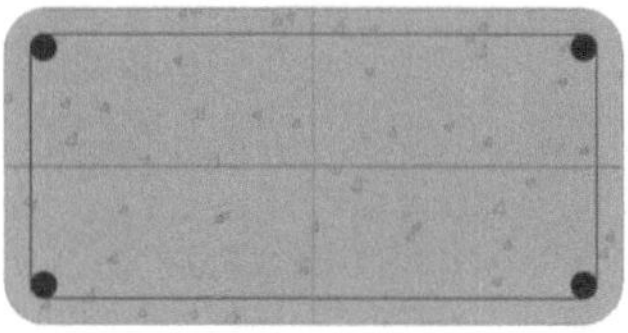

FIGURE 7.P.1

Height	9.84 in.	250 mm
Width	19.69 in.	500 mm
Clear cover	2 in.	51 mm
r_c	0.52 in.	13.21 mm
f'_c	2.66 ksi	18.35 MPa
$n \times t_f$	5 × 0.005 in.	5 × 0.127 mm
E_f	33,350 ksi	230.08 GPa
ε_{fu}	0.015	0.015
f_{yt}	41.47 ksi	286.10 MPa
Tie spacing	7.87 in.	200 mm
Tie diameter	($\phi = 0.31$ in.)	($\phi = 7.87$ mm)
Longitudinal bars	4 ($\phi = 0.71$ in.)	4 ϕ 18.03 mm
f_y	81.2 ksi	560.2 MPa

Plot the approximate confined interaction diagram. The tension-controlled experimental point is at $P_n = 194.625$ kips and $M_n = 90.036$ k-ft. Note that bending is about the y axis ($\alpha = 90°$).

Problem 7.5

One of the square columns (MS6) that was tested by Memon and Sheikh (2005) is investigated in this problem. The following are the properties of the column, see Figure 7.P.2:

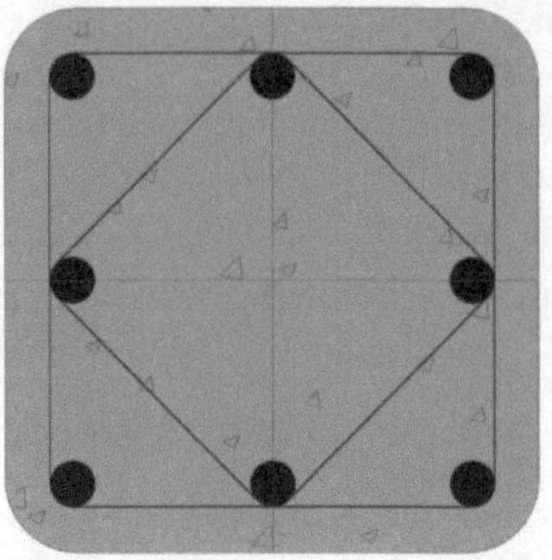

FIGURE 7.P.2

Height	12 in.	305 mm
Width	12 in.	305 mm
Clear cover	1 in.	25 mm
r_c	0.64 in.	16.26 mm
f'_c	6.41 ksi	44.22 MPa
$n \times t_f$	3 × 0.05 in.	3 × 1.27 mm
E_f	2,864.33 ksi	19,761 MPa
ε_{fu}	0.0228	0.0228
f_{yt}	66.625 ksi	459.65 MPa
Tie spacing	11.8 in.	300 mm
Tie diameter	#3	(ϕ = 9.5 mm)
Longitudinal bars	8 (ϕ = 0.77 in.)	8 ϕ 19.56 mm
f_y	67.425 ksi .	465.2 MPa

Plot the approximate confined interaction diagram. The experimental point near balanced failure is at P_n = 569.7 kips and M_n = 211.068 k-ft.

REFERENCES

Abd El Fattah, A. M. 2012. Behavior of concrete columns under various confinement effects. PhD diss., Kansas State University, Manhattan, KS.

Abd El Fattah, A. M., H. A. Rasheed, and A. Esmaeily. 2011. A new eccentricity-based simulation to generate ultimate confined interaction diagrams for circular concrete columns. *Journal of the Franklin Institute-Engineering and Applied Mathematics* 348 (7): 1163–76.

ACI 318-11. 2011. Building code requirements for structural concrete and commentary. ACI Committee 318, Farmington Hills, MI.

ACI 440.2R-08. 2008. Guide for the design and construction of externally bonded FRP systems for strengthening concrete structures. ACI Committee 440, Farmington Hills, MI.

Al-Rahmani, A. H., and H. A. Rasheed. 2014. Concentric and eccentric behavior of rectangular columns confined by steel ties and FRP wrapping. Paper presented at ACI Spring 2014 Convention, March 23–27, Reno, NV.

Bousias, S. T. Triantafillou, M. Fardis, L. Spathis, and B. O'Regan. 2004. Fiber-reinforced polymer retrofitting of rectangular reinforced concrete columns with or without corrosion. *ACI Structural Journal* 101 (4): 512–20.

Carey, S. A., and K. A. Harries. 2005. Axial behavior and modeling of small-, medium- and large-scale circular sections confined with CFRP jackets. *ACI Structural Journal* 102 (4): 596–604.

Chaallal, O., and M. Shahawy. 2000. Performance of fiber-reinforced polymer-wrapped reinforced concrete column under combined axial-flexural loading. *ACI Structural Journal* 97 (4): 659–68.

Darby, A., R. Coonan, T. Ibell, and M. Evernden. 2011. FRP confined square columns under concentric and eccentric loading. In *Proceedings of the 5th international conference on advanced composites in construction 2011*, 264–75. Chesterfield, U.K.: NetComposites.

Eid, R., N. Roy, and P. Paultre. 2009. Normal- and high-strength concrete circular elements wrapped with FRP composites. *ASCE Journal of Comp. for Const.* 13 (2): 113–24.

Kupfer, H.B., H. K. Hilsdorf, and H. Rüsch. 1969. Behaviour of Concrete under Biaxial Stresses, *ACI Journal* 66(8), 656–666.

Lam, L., and J. G. Teng. 2003a. Design-oriented stress–strain model for FRP-confined concrete. *Construction and Building Materials* 17 (6-7): 471–89.

Lam, L., and J. G. Teng. 2003b. Design-oriented stress–strain model for FRP-confined concrete in rectangular columns. *Journal of Reinforced Plastics and Composites* 22 (13): 1149–86.

Mander, J. B. 1983. Seismic design of bridge piers. PhD diss. University of Canterbury, Christchurch, New Zealand.

Mander, J. B., M. J. N. Priestley, and R. Park. 1988. Theoretical stress–strain model for confined concrete. *ASCE Journal of Structural Engineering* 114 (8): 1804–26.

Memon, M., and S. Sheikh. 2005. Seismic resistance of square concrete columns retrofitted with glass fiber-reinforced polymer. *ACI Structural Journal* 102 (5): 774–83.

Pessiki, S., K. A. Harries, J. Kestner, R. Sause, and J. M. Ricles. 2001. The axial behavior of concrete confined with fiber reinforced composite jackets. *ASCE Journal of Composites for Construction* 5 (4): 237–45.

Rasheed, H. A., A. M. Abd El Fattah, A. Esmaeily, J. P. Jones, and K. F. Hurst. 2012. Software for adaptable eccentric analysis of confined concrete circular columns. *Computers and Concrete, an International Journal* 10 (4): 331–47.

Rasheed, H. A., and K. S. Dinno. 1994. An efficient nonlinear analysis of R.C. sections. *Computers and Structures, International Journal* 53 (3): 613–23.

Saadatmanesh, H., M. R. Ehsani, and L. Jin. 1996. Seismic strengthening of circular bridge piers with fiber composites. *ACI Structural Journal* 93 (6): 639–47.

Sheikh, S. A., and G. Yau. 2002. Seismic behaviour of concrete columns confined with steel and fiber-reinforced polymers. *ACI Structural Journal* 99 (1): 72–80.

Willam, K. J., and E. P. Warnke. 1975. Constitutive model for the triaxial behaviour of concrete. In *Proceedings of International Association for Bridge and Structural Engineering*. Vol. 19. Zurich: IABSE.

8 Installation

8.1 OVERVIEW

Installation of an FRP system depends on the manufacturer's procedure, which could vary according to the system. Installation also can vary based on the type and condition of the member and the various environmental factors directly impacting the installation. Application personnel must be trained by the manufacturer or its agent. Deviation from the original procedure developed for a certain system needs approval from the manufacturer prior to its acceptance.

8.2 ENVIRONMENTAL CONDITIONS

Conditions related to temperature, moisture, and humidity during installation affect the FRP system installed. For example, primers, putty, and adhesive resin cannot be applied to cold surfaces. A heat source may be used to heat the surface without contaminating it or the uncured FRP system. However, if the temperature is lower than specified, the resin may not cure properly, and fiber saturation may be inadequate.

Similarly, applying resin to wet or moist surfaces may affect the curing or fiber saturation as well, unless the resin is said to be water or moisture resistant. If moisture gets into the uncured resin, it may create bubbles that could impact the bond between the fibers or with the concrete substrate.

8.3 SURFACE PREPARATION AND REPAIR

This is the main key to the success of the FRP system in performing properly to strengthen concrete members. Debonding or localized separation of the FRP system may take place due to poorly prepared concrete substrate. Detailed guidance should be obtained from the FRP manufacturer about surface preparation. However, this chapter provides generalized guidelines to the proper repair and surface preparation for externally bonded and near-surface-mounted FRP installation.

ACI 440.2R-08 categorized this topic into the following areas:

1. Surface repair

   Any damage to the surface in terms of spalling, breakage of cover material, or unevenness of substrate surface must be repaired prior to the installation of the FRP system. Repair using resin material compatible with the substrate and FRP system is expected to restore a proper and even surface for FRP installation. The compatibility of the resin material with the adhered material and the substrate is critical, as seen

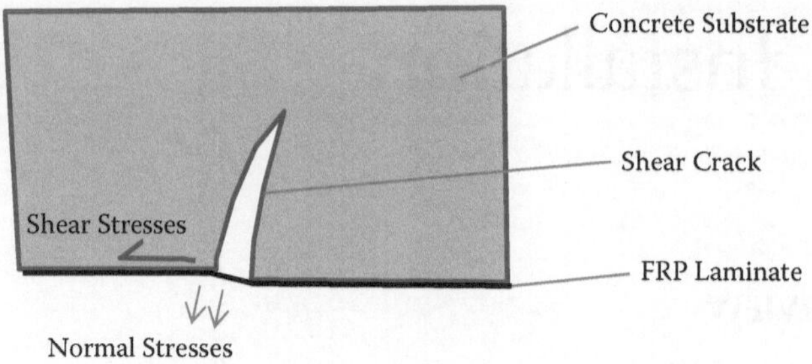

FIGURE 8.1 Intermediate crack-induced FRP debonding.

in a study by Saadatmanesh and Ehsani (1990), which showed that the effectiveness of FRP plates in strengthening beams is highly dependent on the selection of the resin adhesive.

2. Crack injection by resin

   Cracks wider than 0.3 mm (0.01 in.) may negatively impact the behavior of FRP-strengthened members by provoking early premature delaminations or fiber kinking (ACI 440.2R-08), as seen in Figure 8.1. Such cracks need to be epoxy-injected prior to surface preparation according to ACI 224.1R-07. For more detailed step-by-step procedure, refer to Reed et al. (2003, 2005). Narrower cracks in more aggressive environments need to be sealed to prevent corrosion of the main steel reinforcement.

3. Damage caused by corrosion

   When it is evident that the member to be strengthened has undergone corrosion damage, the cover needs to be removed and the corroded bars need to be cleaned. The cover must then be repaired prior to applying FRP materials to avoid putting the integrity of the system in question, as seen in Figure 8.2.

4. Surface preparation

   Whenever the bond between the substrate and FRP system is critical for the load transfer, special attention must be provided to surface preparation in the so-called bond-critical applications like the flexural or shear strengthening of beams, columns, slabs, or walls. This surface preparation is not so critical for the so-called contact-critical applications that have adhesive bonding during installation. A typical application of this category is lateral confinement of columns.

   • Surface preparation for bond-critical application

The surface to which the FRP is to be bonded should be free of weak or loose material and should be recently exposed to the aggregate level by means of grinding or sandblasting (Rasheed et al. 2011; Larson, Peterman, and Rasheed 2005). Grinding typically uses a diamond rotary grinder to grind the surface until it is roughened to the aggregate level, as seen in Figure 8.3.

FIGURE 8.2 Damage caused by corrosion: (a) before repair, (b) after repair prior to strengthening, and (c) after strengthening. (Courtesy of Fyfe Inc.)

Sandblasting involves the use of a fine-graded silica sand and a high-pressure air compressor to remove any dirt or other foreign substance attached to the concrete surface, leaving a clean, etched surface that FRP bonds to easily, as seen in Figure 8.4.

The corners in rectangular concrete cross sections should be rounded to a minimum of 0.5 in. (13 mm) radius when fibers are wrapped around them to avoid any FRP stress concentrations and possible voids between the concrete and the FRP system, as seen in Figure 8.5. Putty should be used to smooth roughened corners. Obstructions need to be removed prior to FRP

FIGURE 8.3 Grinding of concrete substrate surface (left) roughened vs. original concrete surface (right).

FIGURE 8.4 Surface preparations by sandblasting.

application. Inside corners and concave surfaces require detailing to maintain an intact bond, as seen in Figure 8.6. Dust, dirt, oil, existing coatings, and any other bond barrier material must be removed.

Bug holes and other voids need to be filled by putty and evened out prior to FRP installation. Local out-of-plane variations like form lines must not exceed 1/32 in. or 1 mm (ACI 440.2R-08). Such variations may be removed by grinding prior to surface preparation or may be evened out using putty if the variations are small. The concrete surface must be dry if it is to bond to the FRP system.

- Surface preparation for contact-critical application

FIGURE 8.5 Rounded beam corners by grinding.

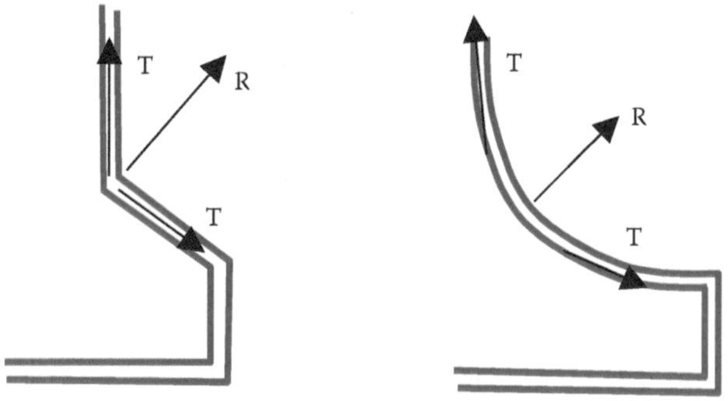

FIGURE 8.6 Inside corners and concave surfaces needing extra detailing.

Surface preparation described for bond-critical applications is not necessary here. However, surface preparation should enable a continuous and effective contact surface. Contact surfaces must be flat or convex to help engage the FRP system during loading. Surface spalls or voids need patching with a compatible repair material. Soft materials covering columns, like gypsum plaster, must be removed prior to applying the FRP system.

- Surface preparation for NSM application

  A diamond saw is typically needed to cut grooves into the concrete cover surface. Existing steel needs to be kept intact during the groove making. The cut grooves should leave sound concrete surfaces and be free of loose materials that prohibit a good bond. The grooves should be dry and should be completely filled with resin. Once the FRP reinforcement is placed into the groove, the extra adhesive resin needs to be scraped off. This resin is specified by the NSM manufacturer.

5. Resin mixing

   Resins are best to be mixed by hand stirring in disposable plastic pots. However, small electric mixing blades may be used for stirring if cleaning them afterwards is not difficult or time consuming. Mixing should be in small quantities to make sure that the resin can be applied within its pot life. Once the pot life is exceeded, the mixed resin should be discarded, since it becomes very viscous to the point that it can no longer effectively saturate the fibers.

   All resin parts should be mixed in the proper ratio prescribed by the manufacturer until the mix is uniform. Resin parts are typically color coded so that proper mixing is indicated by the removal of color streaks. All mixing factors need to be provided by the manufacturer (mixing proportions, methods, time, etc.).

6. Installation of FRP system

   A generic schematic showing the sequence of FRP layers installed is presented in Figure 8.7.

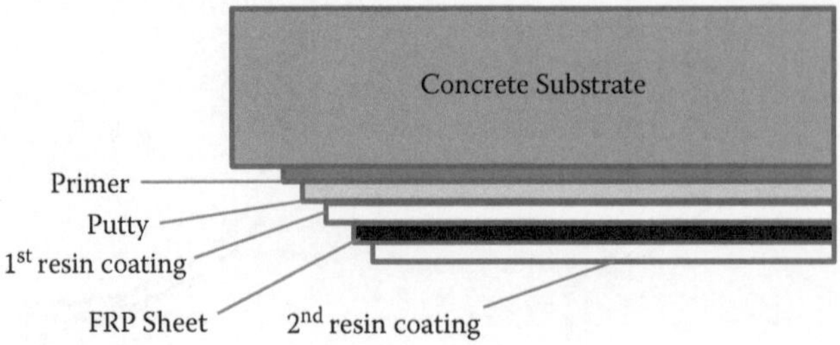

FIGURE 8.7 Wet lay-up process and layers.

a. Primer and resin putty

Primer is the first coat to be applied to the substrate concrete surface using a paint roller. The primer is immediately followed by resin putty that is used to fill in small bug holes on the concrete surface.

b. FRP sheet placement

Before the primer and resin putty are completely allowed to cure, the first layer of the saturating resin is generously applied to the surface using a paint roller, as shown in Figure 8.8. Then a sheet of fibers is pressed into the saturating resin or adhesive with a ribbed aluminum or plastic roller to prevent air pockets from forming under the fiber sheet.

Consecutive layers of saturating resin or adhesive and of fiber sheets should be applied before completely curing the existing saturant. ACI 440.2R-08 recommends inter-layer surface preparation by light surface sanding or using solvent if previous layers are fully cured. The recommendation of the FRP system manufacturer is important.

c. Precured laminate placement

Precured laminates are typically bonded to concrete surfaces using the manufacturer's recommended adhesive. Adhesive should be uniformly applied to the bonding surface to maintain an even interface. The precured laminates themselves should be clean and free of dust and should be pressed against

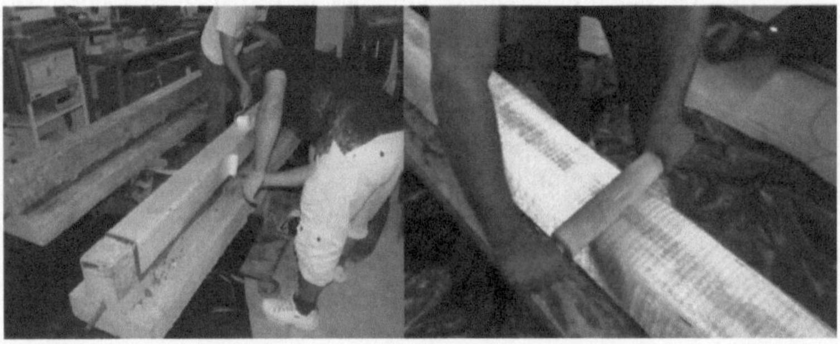

FIGURE 8.8 Applying resin to FRP with a paint roller (left) and clearing the air pockets with a wooden roller (right).

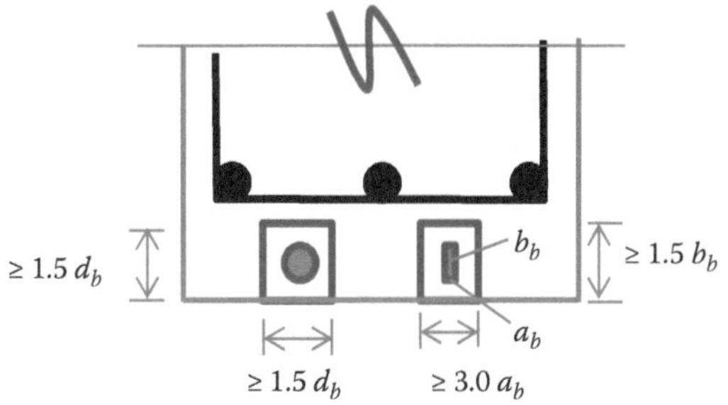

FIGURE 8.9 Minimum dimensions of grooves in NSM application.

the uncured adhesive in such a way as to roll out all entrapped air. Surface preparations described earlier are needed prior to applying the adhesive layer.

d. NSM bar placement

Grooves are cut to place NSM rectangular or circular bars into the cover part of the section. Grooves are dimensioned as shown in Figure 8.9 to allow enough adhesion around the bars. Surface preparations must be performed as described earlier. The manufacturer of the NSM system should be consulted on the type of adhesive to use.

7. Alignment of the FRP sheets

FRP sheets should be properly aligned with the beam axis. Deviation from the intended fiber angle orientation may lead to significant reduction in modulus and strength for angle differences as small as 5° (ACI 440.2R-08). Fabrics and fiber sheets should be carefully placed to maintain fiber straightness. Any kinks or folds in fibers during installation may adversely affect the performance, so these should be reported to the design engineer (ACI 440.2R-08).

8. Curing of resins

This is a temperature-dependent and time-dependent process. For resins cured at room or ambient temperature, curing may take several days for most of the cross linking to take place. For resins cured at higher temperature levels, resin must be kept at a certain temperature for a certain time span. The manufacturer's recommended curing process should be carefully followed without deviations. Curing should be visually inspected, and problems with curing due to expired resins beyond their shelf life should be reported to the designer.

REFERENCES

ACI 224.1R-07. 2007. Causes, evaluation and repair of cracks in concrete structures. ACI Committee 224. Farmington Hills, MI.

ACI 440.2R-08. 2008. Guide for the design and construction of externally bonded FRP systems for strengthening concrete structures. ACI Committee 440. Farmington Hills, MI.

Larson, K. H., R. J. Peterman, and H. A. Rasheed. 2005. Strength-fatigue behavior of FRP strengthened prestressed concrete T beams. *ASCE Journal of Composites for Construction* 9 (4): 313–26.

Rasheed, H. A., M. Nassajy, S. Al-Subaie, S. M. Abrishamchian, and A. Al-Tamimi. 2011. Suppressing delamination failure mode in concrete beams strengthened with short CFRP laminates. *Mechanics of Advanced Materials and Structures* 18 (3): 194–200.

Reed, C. E., R. J. Peterman, and H. A. Rasheed. 2005. Evaluating FRP repair method for cracked prestressed concrete bridge members subjected to repeated loadings: Phase 1. Report No. K-TRAN: KSU-01-2. Kansas Department of Transportation (KTRAN), Topeka.

Reed, C. E., R. J. Peterman, H. A. Rasheed, and D. Meggers. 2003. Adhesive applications used during the repair and strengthening of 30-year-old prestressed concrete girders. *Transportation Research Record*, no. 1827:36–43.

Saadatmanesh, H., and M. R. Ehsani. 1990. FRP plates can strengthen beams. *Concrete International, ACI* 99 (7): 41–47.

Index

A

B

C

D

E

F

G

H

I

K

L

M

N

O

P

R

S